Copyright © 2024 Humbert Cole

All rights reserved. No part of the material protected by this copyright may be reproduced or utilized in any form, electronic or mechanical, including photocopying, recording, or by any information storage and retrieval system, without written permission from the copyright owner.

Contact the author:

- humbertcole3@gmail.com
- facebook.com/HumbertColeMath

Contents

1 **Prime and composite numbers**   **1**
    1.1   Introduction   1
    1.2   Primality test   6
    1.3   Sieve of Eratosthenes   7

2 **Prime factorization**   **13**
    2.1   Euclid's first theorem   13
    2.2   Introduction to tuples   15
    2.3   Fundamental theorem of arithmetic   17
    2.4   Standard form of an integer   20
    2.5   Finding the prime factorization of an integer   22
    2.6   GCD and LCM using prime factorization   23

3 **Infinitude of primes**   **29**
    3.1   Introduction   29
    3.2   Sequence of pairwise coprime integers   30
    3.3   Inequalities of prime numbers   33

4 **Forms of primes**   **37**
    4.1   Mersenne numbers   37
    4.2   Sums and products   39
    4.3   Fermat numbers   42
    4.4   Partition into primes   49
    4.5   Goldbach conjecture   50
    4.6   Forms of primes   56

5 **Primes in AP**   **63**
    5.1   Introduction   63
    5.2   Dirichlet's theorem   70

6 **Prime gaps**   **73**
    6.1   Twin primes   73

    6.2   Large gaps between primes   76
    6.3   Sequence of prime gaps   78
    6.4   Change in length of gap   79

7   Distribution of primes   81

    7.1   Introduction   81
    7.2   Bertrand's postulate   82
    7.3   Prime-counting function   87
    7.4   Scarcity of primes   91
    7.5   Prime number theorem   92

CHAPTER 1

Prime and composite numbers

1.1 Introduction

1. In this section, we classify integers greater than 1 into prime and composite integers. Then we show how to determine if an integer is composite.

2. **Primes and composites.** Each integer greater than 1 has at least two positive factors: 1 and itself. For example 25 has the positive factors 1 and 25.

3. Let n be an integer greater than 1. Each positive divisor of n is not greater than n. There are exactly n positive integers not greater than n. Hence n has at most n positive divisors.

4. Each integer greater than 1 has at least two positive factors. If such an integer has no other positive factors, then we say the integer is a prime number.

5. **Definition of prime number.** A prime number is an integer greater than 1 with exactly two positive factors. Hence, a prime number has no positive factors other than 1 and itself. A prime number is also known as a prime. An example of a prime number is 2. Its only positive factors are 1 and 2. The first few prime numbers are $2, 3, 5, 7, 11$ and 13.

6. Since the only divisors of a prime p are 1 and p it follows that for all integers a
$$\gcd(a, p) = 1 \text{ or } p$$

7. **Definition of composite number.** A composite number is an integer greater than 1 that is not a prime number. Hence a composite number is an integer greater than 1 with more than two positive factors. An example of a composite number is 4. It has more than two positive factors: $1, 2, 4$. The first few composite numbers are $4, 6, 8, 9, 10$.

8   Let n be an integer greater than 1. Then n is composite if and only if there exists an integer d such that $d \mid n$ and $1 < d < n$.

Proof

Part 1: If n is composite then there exists an integer d such that $d \mid n$ and $1 < d < n$ Let n be a composite integer. Thus n is not prime and there exists some positive divisor d of n other than 1 and n. Since $d \mid n$, it follows that $d \leq n$. Since $d \neq n$, it follows that $d < n$. Since d is a positive integer other than 1, $d > 1$. Thus $1 < d$ and $d < n$. Hence $1 < d < n$.

Part 2: If there exists an integer d such that $d \mid n$ and $1 < d < n$ then n is composite   Let n be an integer greater than 1 such that there exists a positive divisor d of n such that $1 < d < n$. Thus $d \mid n$, $d \neq 1$ and $d \neq n$. Hence n has a positive divisor other than 1 and n. Hence n is not prime. Thus n is composite. ∎

9   The integer 2 is the only even prime

Proof   Assume $p > 2$ is prime and p is even. Thus $2 \mid p$ and $1 < 2 < p$. Hence p is composite. This contradicts with the assumption. Hence the assumption must be wrong. Thus 2 is the only even prime. ∎

10   Let a, p be integers with p prime. Then $p \nmid a$ if and only if $\gcd(a, p) = 1$.

Proof   We know that
$$\gcd(a, p) = 1 \text{ or } p$$
Since $p \nmid a$, it follows that $\gcd(a, p) \neq p$. Thus
$$\gcd(a, p) = 1$$
Conversely we have that $\gcd(a, p) = 1$. Assume $p \mid a$. Then $p \mid a$ and $p \mid p$. Thus $p \mid \gcd(a, p)$. Hence $p \mid 1$. Thus $p \leq 1$. This is absurd because each prime is greater than 1. Hence the assumption must be wrong. Thus
$$p \nmid a$$

∎

11 **How to determine if an integer is composite.** Here we show that for each integer n greater than 1, n is composite if and only if there exists a prime divisor p of n such that $p \leq \sqrt{n}$. Before we prove this, we prove some lemmas and corollaries.

12 Each integer greater than 1 has a prime divisor.

   Proof Let n be an integer greater than 1. Let A be the set of divisors d of n such that $d > 1$. By the well-ordering property, A has a least element, say p. We claim that p is prime. Assume p is not prime. Thus p is composite. Hence there exists some integer d such that $1 < d < p$ and $d \mid p$. Hence $d \mid p$ and $p \mid n$. It follows that $d \mid n$. Hence $d \in A$. Since $d < p$, it follows that d is an element of A that is smaller than p. This is absurd because p is the smallest element of A. It follows that p is prime. Hence n has a prime divisor. Therefore each integer greater than 1 has a prime divisor. ■

13 Let a, b be integers. Then $\gcd(a, b) = 1$ if and only if a and b have no prime factor in common.

   Proof

   Part 1: If $\gcd(a, b) = 1$ then a and b have no prime factor in common
   Since $\gcd(a, b) = 1$ it follows that no common factor of a and b is greater than 1. Each prime is greater than 1. Hence a and b have no prime factor in common.

   Part 2: If a and b have no prime factor in common then $\gcd(a, b) = 1$
   Let d be a positive common factor of a and b. Assume $d \neq 1$. Hence $d > 1$ Thus there exists a prime p such that $p \mid d$. Hence p is a common prime factor of a and b. This is absurd. Hence $d = 1$. Thus each positive common factor of a and b is equal to 1. Since $\gcd(a, b)$ is a positive common factor of a and b, it follows that $\gcd(a, b) = 1$. ■

14 For all positive real numbers a, b, if $a > \sqrt{ab}$ then $b < \sqrt{ab}$.

   Proof We are given that
   $$a > \sqrt{ab}$$

$$\sqrt{a}\sqrt{a} > \sqrt{a}\sqrt{b}$$

Divide both sides by $\sqrt{a}$

$$\sqrt{a} > \sqrt{b}$$

Multiply both sides by $\sqrt{b}$

$$\sqrt{a}\sqrt{b} > \sqrt{b}\sqrt{b}$$
$$\sqrt{ab} > b$$
$$b < \sqrt{ab}$$

■

15 n is composite if and only if there exists an integer d such that $d \mid n$ and $1 < d \leq \sqrt{n}$

Proof

Part 1: If n is composite then there exists an integer d such that $d \mid n$ and $1 < d \leq \sqrt{n}$ Let n be a composite integer. Thus there exists some integer c such that $c \mid n$ and $1 < c < n$. It is either $c \leq \sqrt{n}$ or $c > \sqrt{n}$. In the first case, $c \leq \sqrt{n}$. Let $d = c$. Hence $d \mid n$ and $1 < d \leq \sqrt{n}$.

In the second case, $c > \sqrt{n}$. Since $c \mid n$, it follows that $n = cd$ for some integer d such that $1 < d \leq \sqrt{n}$. Hence, $d \mid n$ and $1 < d \leq \sqrt{n}$.

Thus in both cases there exists an integer d such that $d \mid n$ and $1 < d \leq \sqrt{n}$.

Part 2: If there exists an integer d such that $d \mid n$ and $1 < d \leq \sqrt{n}$ then n is composite Let n be an integer greater than 1 such that there exists a positive divisor of n such that $1 < d \leq \sqrt{n}$. Since $n > 1$, it follows that $\sqrt{n} < n$. Hence, $d \leq \sqrt{n}$ and $\sqrt{n} < n$. Thus $d < n$. Hence $1 < d < n$. Thus $d \mid n$ and $1 < d < n$. It follows that n is composite.

■

16 n is composite if and only if there exists a prime divisor p of n such that $p \leq \sqrt{n}$.

Proof

Part 1: If n is composite then there exists a prime divisor p of n such that $p \leq \sqrt{n}$ Let n be a composite integer. Thus there exists some integer d such that $d \mid n$ and $1 < d \leq \sqrt{n}$. Since $d > 1$, it follows that there exists a prime divisor p of d. Thus $p \leq d$. Hence $p \leq d$ and $d \leq \sqrt{n}$. It follows that $p \leq \sqrt{n}$.

Part 2: If there exists a prime divisor p of n such that $p \leq \sqrt{n}$ then n is composite Let n be an integer greater than 1 such that there exists a prime divisor p of n such that $p \leq \sqrt{n}$. Since $p > 1$, it follows that $1 < p \leq \sqrt{n}$. By a previous theorem, it follows that n is composite.

∎

Exercises 1.1

1. Prove the following.

   a Product of two prime numbers is composite.

   b Product of two composite numbers is composite.

2. Let n be an integer greater than 1. Let p be a prime number. Prove the following.

   a n is a composite integer if and only if n is the product of two integers greater than 1.

   b If $p \nmid n$ for all primes p such that $p \leq \sqrt[3]{n}$, then n is either a prime or a product of two primes.

   (Hint: Use contradiction to show that n cannot have three distinct primes.)

3. Let n be an integer greater than 1. Prove the following.

   a The smallest divisor of n that is greater than 1 is a prime.

   b n is composite if and only if n is the sum of two noncoprime integers.

4. Let n be an integer greater than 1. Let a, b, c be positive integers. Prove the following.

   a If n is prime and $n = a + b + c$ then $\gcd(a, b, c) = 1$.

   b If $\gcd(a, b) > 1$ then $a + b$ is composite.

c If $\gcd(a,b) > 1$ then $ax + by$ is composite for all positive integers x, y. In other words if a and b are not coprime then every of their linear combination using positive integers is composite.

5 Let n be an integer greater than 1. Let a, b be integers. Prove the following.

a If n is a composite integer which is not the square of a prime then there exist integers a, b such that

$$n = ab$$
$$1 < a < b < n$$

(Hint: Let a be the smallest prime factor of n. Thus $n = ab$ for some integer b. Use contradiction to show that $b > a$.)

b If $1 \leq a \leq n$, $1 \leq b \leq n$ and $a \neq b$ then $a \cdot b \mid n!$

c If $n > 4$ and n is composite then $n \mid (n-1)!$

(Hint: Consider two cases: when n is a perfect square and when it is not.)

1.2 Primality test

1 In this section, we discuss how to check whether an integer greater than 1 is prime.

2 Consider the integer 131. This integer is composite if and only if we can find a prime less than or equal to $\sqrt{131}$ that divides it.

3 The primes not greater than $\sqrt{131}$ are 2, 3, 5, 7, 11. Since none of these primes divides 131, it follows that 131 is not composite. Thus 131 is a prime number.

4 As another example, consider the integer 121. The primes not greater than $\sqrt{121}$ are 2, 3, 5, 7, 11. Since 11 divides 121, it follows that 121 is composite.

Exercises 1.2

Check whether each integer is prime.

1 1463

2 9537

3   557

4   1553

5   5303

6   6709

1.3 Sieve of Eratosthenes

1   In this section, we describe the sieve of Eratosthenes and consider some examples. Then we explain with a proof why this method works.

2   **Description of the sieve of Eratosthenes.** Consider the problem of finding the primes in the list $2, \ldots, 100$. This is equivalent to finding the composite numbers in the list, since we can just exclude the composite numbers to get the prime numbers.

3   The sieve of Eratosthenes is a method of finding the prime numbers in a list by excluding the composite integers in the list.

4   This is how the sieve of Eratosthenes works: Consider a list of integers. We want to find all the primes in the list.

Step 1. Determine the largest integer in the list, say N.

Step 2. Find the primes $p_1, \ldots, p_m$ which are not greater than $\sqrt{N}$.

Step 3. Exclude all the multiples of p_1, apart from p_1, in the list. Do this for all the primes $p_1, \ldots, p_m$. The integers left in the list are primes.

5   Let us consider some examples to illustrate this method.

6   **Example 1.** Let us find the primes in the list $2, 3, \ldots, 36$.

Solution The list can be written as an array:

$$\begin{array}{ccccccc} 2 & 3 & 4 & 5 & 6 & 7 & 8 \\ 9 & 10 & 11 & 12 & 13 & 14 & 15 \\ 16 & 17 & 18 & 19 & 20 & 21 & 22 \\ 23 & 24 & 25 & 26 & 27 & 28 & 29 \\ 30 & 31 & 32 & 33 & 34 & 35 & 36 \end{array}$$

The largest integer in the list is 36. The primes not greater than $\sqrt{36}$ are $2, 3, 5, 7$.

Chapter 1 Prime and composite numbers

If we cancel the multiples of 2, apart from 2, we get

```
 2   3   4̸   5   6̸   7   8̸
 9  1̸0̸  11  1̸2̸  13  1̸4̸  15
1̸6̸  17  1̸8̸  19  2̸0̸  21  2̸2̸
23  2̸4̸  25  2̸6̸  27  2̸8̸  29
3̸0̸  31  3̸2̸  33  34  35  3̸6̸
```

If we cancel the multiples of 3, apart from 3, we get

```
 2   3   4̸   5   6̸   7   8̸
 9̸  1̸0̸  11  1̸2̸  13  1̸4̸  1̸5̸
1̸6̸  17  1̸8̸  19  2̸0̸  2̸1̸  2̸2̸
23  2̸4̸  25  2̸6̸  2̸7̸  2̸8̸  29
3̸0̸  31  3̸2̸  3̸3̸  34  35  3̸6̸
```

If we cancel the multiples of 5, apart from 5, we get

```
 2   3   4̸   5   6̸   7   8̸
 9̸  1̸0̸  11  1̸2̸  13  1̸4̸  1̸5̸
1̸6̸  17  1̸8̸  19  2̸0̸  2̸1̸  2̸2̸
23  2̸4̸  2̸5̸  2̸6̸  2̸7̸  2̸8̸  29
3̸0̸  31  3̸2̸  3̸3̸  34  3̸5̸  3̸6̸
```

If we cancel the multiples of 7, apart from 7, we get

```
 2   3   4̸   5   6̸   7   8̸
 9̸  1̸0̸  11  1̸2̸  13  1̸4̸  1̸5̸
1̸6̸  17  1̸8̸  19  2̸0̸  2̸1̸  2̸2̸
23  2̸4̸  2̸5̸  2̸6̸  2̸7̸  2̸8̸  29
3̸0̸  31  3̸2̸  3̸3̸  34  3̸5̸  3̸6̸
```

The numbers that have not been cancelled are the primes in the list. They are 2, 3, 5, 7, 11, 13, 17, 19, 23, 29, 31.

Example 2. Let us find the primes in the following list.

```
50  51  52  53  54
55  56  57  58  59
60  61  62  63  64
65  66  67  68  69
70  71  72  73  74
```

Solution The largest integer in the list is 74. The primes not greater than $\sqrt{74}$ are $2, 3, 5, 7$.

If we cancel the multiples of 2, apart from 2, we get

$$\begin{array}{ccccc} \cancel{50} & 51 & \cancel{52} & 53 & \cancel{54} \\ 55 & \cancel{56} & 57 & \cancel{58} & 59 \\ \cancel{60} & 61 & \cancel{62} & 63 & \cancel{64} \\ 65 & \cancel{66} & 67 & \cancel{68} & 69 \\ \cancel{70} & 71 & \cancel{72} & 73 & \cancel{74} \end{array}$$

If we cancel the multiples of 3, apart from 3, we get

$$\begin{array}{ccccc} \cancel{50} & \cancel{51} & \cancel{52} & 53 & \cancel{54} \\ 55 & \cancel{56} & \cancel{57} & \cancel{58} & 59 \\ \cancel{60} & 61 & \cancel{62} & \cancel{63} & \cancel{64} \\ 65 & \cancel{66} & 67 & \cancel{68} & \cancel{69} \\ \cancel{70} & 71 & \cancel{72} & 73 & \cancel{74} \end{array}$$

If we cancel the multiples of 5, apart from 5, we get

$$\begin{array}{ccccc} \cancel{50} & \cancel{51} & \cancel{52} & 53 & \cancel{54} \\ \cancel{55} & \cancel{56} & \cancel{57} & \cancel{58} & 59 \\ \cancel{60} & 61 & \cancel{62} & \cancel{63} & \cancel{64} \\ \cancel{65} & \cancel{66} & 67 & \cancel{68} & \cancel{69} \\ \cancel{70} & 71 & \cancel{72} & 73 & \cancel{74} \end{array}$$

If we cancel the multiples of 7, apart from 7, we get

$$\begin{array}{ccccc} \cancel{50} & \cancel{51} & \cancel{52} & 53 & \cancel{54} \\ \cancel{55} & \cancel{56} & \cancel{57} & \cancel{58} & 59 \\ \cancel{60} & 61 & \cancel{62} & \cancel{63} & \cancel{64} \\ \cancel{65} & \cancel{66} & 67 & \cancel{68} & \cancel{69} \\ \cancel{70} & 71 & \cancel{72} & 73 & \cancel{74} \end{array}$$

The numbers that have not been cancelled are the primes in the list. They are $53, 59, 61, 71, 73$.

Example 3. Let us find the primes in the following list.

$$\begin{array}{cccc} 80 & 81 & 3 & 5 \\ 14 & 12 & 32 & 50 \\ 22 & 11 & 15 & 17 \\ 8 & 9 & 78 & 51 \end{array}$$

Solution The largest integer in the list is 81. The primes not greater than $\sqrt{81}$ are $2, 3, 5, 7$.

If we cancel the multiples of 2, apart from 2, we get

~~80~~ 81   3   5
~~14~~ ~~12~~ ~~32~~ ~~50~~
~~22~~ 11 ~~15~~ 17
~~8~~   9   ~~78~~ 51

If we cancel the multiples of 3, apart from 3, we get

~~80~~ 81   3   5
~~14~~ ~~12~~ ~~32~~ ~~50~~
~~22~~ 11 ~~15~~ 17
~~8~~ ~~9~~ ~~78~~ ~~51~~

If we cancel the multiples of 5, apart from 5, we get

~~80~~ 81   3   5
~~14~~ ~~12~~ ~~32~~ ~~50~~
~~22~~ 11 ~~15~~ 17
~~8~~ ~~9~~ ~~78~~ ~~51~~

If we cancel the multiples of 7, apart from 7, we get

~~80~~ 81   3   5
~~14~~ ~~12~~ ~~32~~ ~~50~~
~~22~~ 11 ~~15~~ 17
~~8~~ ~~9~~ ~~78~~ ~~51~~

The numbers that have not been cancelled are the primes in the list. They are $81, 3, 5, 11, 17$.

9 **Why the sieve of Eratosthenes works.** Here, we explain why the sieve of Eratosthenes works.

10 Let n be an integer in the list $2, \ldots, 100$. By a previous theorem, we know that n is composite if and only if we can find a prime p such that $p \mid n$ and $p \leq \sqrt{n}$. Each prime that is not greater than $\sqrt{n}$ is also not greater than $\sqrt{100}$. Hence n is composite if and only if we can find a prime p such that $p \mid n$ and $p \leq \sqrt{100}$.

11 This also works for integers other than 100 as we prove below.

12 Let N, n be integers greater than 1 and let $N > n$. Then n is composite if and only if there exists a prime divisor p of n such that $p \neq n$ and $p \leq \sqrt{N}$.

Proof

1.3 Sieve of Eratosthenes

Part 1: If n is composite then there exists a prime divisor p of n such that $p \neq n$ and $p \leq \sqrt{N}$ We are given that n is composite. Hence there exists a prime divisor p of n such that $p \leq \sqrt{n}$. Since $n < N$, it follows that $\sqrt{n} < \sqrt{N}$. Hence, $p \leq \sqrt{N}$. Also, since n is composite, $n \neq p$. Thus $p \neq n$ and $p \leq \sqrt{N}$.

Part 2: If there exists a prime divisor p of n such that $p \neq n$ and $p \leq \sqrt{N}$ then n is composite Let p be a prime divisor of n such that $p \neq n$. Thus there exists a divisor d of n such that $1 < d < n$. Hence n is composite. ∎

Exercises 1.3

Use the sieve of Eratosthenes to find the following.

1 The primes not greater than 200.

2 The primes between 200 and 300 inclusive.

3 Each prime in the following set.

$$\begin{array}{ccccc} 47 & 13 & 12 & 17 & 51 \\ 26 & 28 & 11 & 31 & 32 \\ 76 & 78 & 81 & 91 & 93 \end{array}$$

CHAPTER 2
Prime factorization

2.1 Euclid's first theorem

1. In this section we prove Euclid's first theorem and one of its corollaries. Then we extend the theorem to more than two integers. We also prove that if each integer in a list of pairwise coprime integers divides an integer then their product divides the integer.

2. **Introduction.** Here, we prove Euclid's first theorem: If p is prime and $p \mid ab$ then $p \mid a$ or $p \mid b$. Then we prove a corollary of this theorem: If $p \mid a^n$ then $p^n \mid a^n$.

3. If p is prime and $p \mid ab$ then $p \mid a$ or $p \mid b$.

   (Euclid's first theorem)

   Proof Since p is prime, it follows that $\gcd(p, a) = 1$ or $\gcd(p, a) = p$. In the first case, $\gcd(p, a) = 1$. By Euclid's lemma, $p \mid b$.

   In the second case, $\gcd(p, a) = p$. Thus p is a common divisor of a and p. Hence $p \mid a$. In each case $p \mid a$ or $p \mid b$. ■

4. If $p \mid a^n$ then $p^n \mid a^n$

   Proof We are given that
   $$p \mid a^n$$
   from Euclid's first theorem we have that
   $$p \mid a$$
   thus
   $$a = pk \quad \text{for some integer } k$$

raise both sides to the nth power

$$a^n = (pk)^n$$
$$= p^n k^n$$
$$a^n = p^n j \quad \text{for some integer } j$$

thus we have that

$$p^n \mid a^n$$

5 **Extension of Euclid's first theorem.** If a prime divides a product of integers then it divides one of the integers in the product.

Proof We prove this theorem by induction. Let n be the number of integers in the product. The statement is true at $n = 1$. If a prime p divides a product of one integer then it divides the integer. Hence p divides one of the integers in the product.

Assume the statement is true at $n = k$. Hence if a prime p divides a product of k integers then p divides one of the integers in the product. Suppose a prime p divides a product of $k + 1$ integers. Let the integers be $a_1, a_2, \ldots, a_k, a_{k+1}$. Thus $p \mid a_1 \cdots a_{k+1}$. By Euclid's first theorem, $p \mid a_1 \cdots a_k$ or $p \mid a_{k+1}$. By the induction hypothesis, $p \mid a_1 \cdots a_k$ implies p divides one of $a_1, \ldots, a_k$. Hence p divides one of $a_1, \ldots, a_k$ or $p \mid a_{k+1}$. Thus p divides one of $a_1, \ldots, a_k, a_{k+1}$. Thus the statement is true for $n = k + 1$. Therefore the statement is true. ■

6 **Relatively prime integers and Euclid's first theorem.** Here we show that if $a \mid n$, $b \mid n$ and $\gcd(a, b) = 1$ then $ab \mid n$. Then we extend this theorem to more than two pairwise coprime integers.

7 For all integers a, b, n, if $a \mid n$, $b \mid n$ and $\gcd(a, b) = 1$ then $ab \mid n$.

Proof Since $a \mid n$ it follows that

$$n = ak \tag{1}$$

for some integer k. Since $b \mid n$ and $n = ak$, we may write $b \mid ak$. Since $\gcd(a, b) = 1$, by Euclid's lemma, $b \mid k$. Hence $k = bj$ for some integer j. Hence we may write (1) as

$$n = a(bj)$$
$$= ab \cdot j$$

Thus $ab \mid n$. ■

8   Let $a_1, \ldots, a_m$ be pairwise coprime integers where m is a positive integer. If $a_1 \mid n, \ldots, a_m \mid n$ then $a_1 \cdots a_m \mid n$.

Proof We prove the statement by induction on m. It is true at $m = 1$. If $a_1 \mid n$ then $a_1 \mid n$. Assume it is true for $m = k$. Thus if $a_1 \mid n, \ldots, a_k \mid n$ then $a_1 \cdots a_k \mid n$. Let $a_1 \mid n, \ldots, a_k \mid n, a_{k+1} \mid n$. Thus $a_1 \cdots a_k \mid n$ and $a_{k+1} \mid n$. Since the integers $a_1, \ldots, a_k, a_{k+1}$ are pairwise coprime, it follows that
$$\gcd(a_1 \cdots a_k, a_{k+1}) = 1.$$
Hence by a previous theorem, $a_1 \cdots a_k a_{k+1} \mid n$. Thus the statement is true for $m = k + 1$. ∎

2.2 Introduction to tuples

1   In this section, we discuss tuples of integers. This concept will be used to formalize factorization in another section.

2   **Definition of tuple.** An ordered list of integers is known as a tuple of integers. We write tuples of integers by writing the integers separated by commas and surrounded by parenthesis. An example is $(1, 2)$. Other examples are given below:

- $(7, 2)$
- $(1, 2, 3, 4)$
- $(-10, 5, 7)$

3   A tuple is an ordered list of elements of a set.

4   **Elements of a tuple.** The elements of a tuple are the items that make up the list. For example, the elements of $(5, 6, 8)$ are 5, 6 and 8.

5   **Difference between tuple and set.** Below we list some important differences between tuple and set.

- Tuple can have repetitions. Set has no repetitions. For example,
$$(2, 2, 5, 6) \neq (2, 5, 6)$$
$$\{2, 2, 5, 6\} = \{2, 5, 6\}$$

- Tuple has order. Set has no order. For example,
$$(2, 5, 6) \neq (5, 2, 6)$$
$$\{2, 5, 6\} = \{5, 2, 6\}$$

- Tuple is always finite. Some sets are infinite.

6 **Multiplicity of an element.** The multiplicity of an element in a tuple is the number of times the element appears in the tuple. For example, the multiplicity of 2 in $(2,7,2,2)$ is 3. However, the multiplicity of 5 is zero, since it does not appear in the tuple.

7 **Equality of tuples.** Two tuples are equal if and only if they have the same length and contain the same elements in the same order. That is, two tuples $(a_1,\ldots,a_n)$ and $(b_1,\ldots,b_m)$ are equal if and only if $n = m$ and $a_1 = b_1,\ldots,a_n = b_n$.

8 **Rearrangement of a tuple.** Given a tuple $(2,5,5,3)$, we can form another tuple by changing the order of its elements: $(2,5,3,5)$. This new tuple is called a rearrangement of the original tuple.

9 Observe that the multiplicity of 5 is the same in the tuples $(2,5,5,3)$ and $(2,5,3,5)$. Similarly the multiplicity of 6 is the same in the tuples: it is zero. In general, for each integer n, the multiplicity of n is the same in both tuples.

10 Two tuples A, B of integers are rearrangements of each other if and only if for each integer n the multiplicity of n in A is the same as the multiplicity of n in B.

11 As another example, some rearrangements of the tuple $(1,7,3,1)$ are shown below:

- $(1,7,1,3)$
- $(1,1,7,3)$
- $(7,1,3,1)$

12 **Factorization.** A factorization of an integer is a tuple of integers such that the product of the elements of the tuple yield the integer.

13 For example, the integer 100 has a factorization $(50,2)$ because $50 \times 2 = 100$. Other factorizations of 100 are given below.

- (100)
- $(2,2,25)$
- $(5,5,4)$

- (1, 100)
- (100, 1)
- (2, 50)

14 **Rearrangement of a factorization.** Consider the following factorizations of 100: (50, 2) and (2, 50). These tuples are rearrangements of each other. Thus we say the factorizations are rearrangements of each other.

15 As another example, the following factorizations of 100 are all rearrangements of one another.

- (2, 2, 25)
- (2, 25, 2)
- (25, 2, 2)

16 **Product form of a factorization.** Given a factorization of an integer as a tuple, we can write the product and vice versa. For example, since (2, 25, 2) is a factorization of 100, we know that $100 = 2 \times 25 \times 2$. On the other hand, since $100 = 4 \times 25$, it follows that (4, 25) is a factorization of 100.

17 **Prime factorization.** For each integer n greater than 1, a prime factorization of n is a factorization where all the elements are prime numbers. For example, (2, 5, 2, 5) is a prime factorization of 100 because $100 = 2 \times 5 \times 2 \times 5$ and each element of (2, 5, 2, 5) is a prime.

2.3 Fundamental theorem of arithmetic

1 Consider the integer 60. We may write it as a product of primes:

$$60 = 2 \cdot 2 \cdot 3 \cdot 5$$

2 Thus (2, 2, 3, 5) is a prime factorization of 60. Every other prime factorization of 60 is a rearrangement of (2, 2, 3, 5). For example, the tuples (2, 3, 5, 2) and (3, 2, 2, 5) are prime factorizations of 60, and they are rearrangements of (2, 2, 3, 5).

3 The fundamental theorem of arithmetic states that each integer greater than 1 can be written as a product of primes. Moreover this representation as a product of primes is unique, apart from the order in which the factors occur.

4 We can restate this theorem using the concept of tuples: Each integer greater than 1 has a prime factorization and all different prime factorizations of the integer are rearrangements of one another.

5 Let us break this statement into existence and uniqueness statements: For each integer n greater than 1,

- n has a prime factorization
- all different prime factorizations of n are rearrangements of one another

6 We prove each statement below.

7 **Existence of prime factorization.** Before proving the existence of prime factorization, we prove that each integer greater than 1 is the product of one or more primes.

8 Every integer greater than 1 is either a prime or a product of primes.

Proof We proceed to prove the claim by strong induction. The base case is $n = 2$. We have that 2 is prime. Now assume it is true for $n = 2, 3, \ldots, k$. It is either $k + 1$ is prime or composite. If $k + 1$ is prime then the statement is true at $n = k + 1$. If $k + 1$ is composite then there exist integers a, b such that $k + 1 = ab$ and $1 < a < k + 1$ and $1 < b < k + 1$. Since a, b are among the integers $2, \ldots, k$, we know that a and b are each a product of primes. Thus,

$$a = p_1 p_2 \cdots p_m \qquad (1)$$
$$b = q_1 q_2 \cdots q_r \qquad (2)$$

for some positive integers m, r and for some primes $p_1, \ldots, p_m, q_1, \ldots, q_r$. Now multiply (1) and (2) together

$$ab = (p_1 p_2 \cdots p_m)(q_1 q_2 \cdots q_r)$$
$$k + 1 = (p_1 p_2 \cdots p_m)(q_1 q_2 \cdots q_r)$$

Hence we have written $k + 1$ as a product of primes. Thus the statement is true at $n = k + 1$. Therefore the statement is true for all $n > 1$.

∎

9 Each integer greater than 1 has a prime factorization.

2.3 Fundamental theorem of arithmetic

Proof Let n be an integer greater than 1. Thus n can be written as a product of primes. Hence,
$$n = p_1 \cdots p_m$$
for some primes $p_1, \ldots, p_m$. Thus $(p_1, \ldots, p_m)$ is a prime factorization of n. Hence n has a prime factorization. ∎

10 **Uniqueness of prime factorization apart from order.** Here, we show that every two factorizations of an integer greater than 1 are rearrangements of each other. Thus they contain each prime the same number of times.

11 Every two prime factorizations of an integer greater than 1 are rearrangements of each other.

Proof Let $n > 1$. Let A, B be prime factorizations of n and let p be a prime number. It is sufficient to show that the multiplicity of p in A is same as the multiplicity of p in B.

Let a, b be the multiplicities of p in A, B respectively. Thus p^a is the product of all the p's in A. Let k be the product of all the other integers in A. Thus

$$n = p^a \cdot k \tag{1}$$

and $p \nmid k$. Thus $\gcd(p, k) = 1$. Hence

$$\gcd(p^b, k) = 1 \tag{2}$$

Similarly, p^b is the product of all the p's in B. Let j be the product of all the other integers in B. Thus

$$n = p^b \cdot j \tag{3}$$

and $p \nmid j$. Thus $\gcd(p, k) = 1$. Hence

$$\gcd(p^a, j) = 1 \tag{4}$$

From (3), we know that
$$p^b \mid n$$

From (1), we may write
$$p^b \mid p^a \cdot k$$

Since $\gcd(p^b, k) = 1$,

$$p^b \mid p^a \tag{5}$$

From (1), we know that

$$p^a \mid n$$

From (3), we may write

$$p^a \mid p^b \cdot j$$

Since $\gcd(p^a, j) = 1$,

$$p^a \mid p^b \qquad (6)$$

From (5) and (6), we get $p^a = p^b$. Since $p > 1$, it follows that $a = b$. Thus the multiplicity of p in A is same as the multiplicity of p in B. Therefore A and B are rearrangements of each other. ∎

Exercises 2.3

Let n be an integer greater than 1. Show that

1 $n = 2^k(2m + 1)$

for some nonnegative integers m, k.

2 $n = 3^k(3m \pm 1)$

for some nonnegative integers m, k

2.4 Standard form of an integer

1 In this section, we prove that each integer greater than 1 has a unique standard form. Then we discuss how to convert standard form to prime factorization and vice versa.

2 **Definition of standard form.** Consider the integer 2646 we can write it as a product of distinct prime powers:

$$2646 = 2 \times 3^3 \times 7^2$$

We say $2 \times 3^3 \times 7^2$ is the standard form of 2646.

3 Let n be an integer greater than 1. If $n = p_1^{k_1} \cdots p_m^{k_m}$ for some primes $p_1 < \cdots < p_m$ and positive integers $k_1, \ldots, k_m$, then the expression $p_1^{k_1} \cdots p_m^{k_m}$ is known as the standard form of n.

2.4 Standard form of an integer

4 **Existence of standard form.** Each integer greater than 1 has a standard form.

Proof Let n be an integer greater than 1. Let $p_1, \ldots, p_m$ be the prime factors of n in ascending order. By the fundamental theorem of arithmetic, n has a prime factorization, say A. Each integer in A is a prime factor of n and each prime factor of n is in A. Hence the product of all integers in A can be written as $p_1^{k_1} \cdots p_m^{k_m}$ where $k_1, \ldots, k_m$ are the multiplicities of $p_1, \ldots, p_m$ respectively in A. Hence $n = p_1^{k_1} \cdots p_m^{k_m}$. Each prime factor of n appears at least once in A. Hence the integers $k_1, \ldots, k_m$ are positive integers. Thus

$$n = p_1^{k_1} \cdots p_m^{k_m}$$

for some prime numbers $p_1 < \cdots < p_m$ and for some positive integers $k_1, \ldots, k_m$. Thus $p_1^{k_1} \cdots p_m^{k_m}$ is a standard form of n. ∎

5 **Uniqueness of standard form.** Each integer has at most one standard form

Proof Let $n > 1$. We prove the statement by considering two arbitrary standard forms of n and then showing that they are equal.

Let $p_1^{k_1} \cdots p_m^{k_m}$ be a standard form of n. Thus $n = p_1^{k_1} \cdots p_m^{k_m}$. Let A be a tuple such that $p_1, \ldots, p_m$ have the multiplicities $k_1, \ldots, k_r$ respectively in A and A contains no other elements.

Let $q_1^{j_1} \cdots q_r^{j_r}$ be another standard form of n. Thus $n = q_1^{j_1} \cdots q_r^{j_r}$. Let B be a tuple such that $q_1, \ldots, q_r$ have the multiplicities $j_1, \ldots, j_r$ respectively in B and B contains no other elements.

Thus A and B are prime factorizations of n. Hence B is a rearrangement of A. Hence the sets $\{p_1, \ldots, p_m\}$ and $\{q_1, \ldots, q_r\}$ are equal.

Since $q_1, \ldots, q_r$ are distinct integers and $p_1, \ldots, p_m$ are distinct integers, it follows that $m = r$. Since there is only one way of arranging the elements of a set in ascending order, it follows that $q_1 = p_1, q_2 = p_2, \ldots, q_m = p_m$. Now j_1 is the multiplicity of p_1 in B. Since A and B are rearrangements, it follows that the multiplicities of p_1 in A and B are equal. Thus $j_1 = k_1$. Similarly, $j_2 = k_2, \ldots, j_m = k_m$. Thus, the standard form $q_1^{j_1} \cdots q_r^{j_r}$ is equal to the standard form $p_1^{k_1} \cdots p_m^{k_m}$. ∎

6 **Prime factorization and standard form.** We can easily convert the standard form of an integer to a prime factorization and vice versa.

22   Chapter 2 Prime factorization

7   For example, consider the integer 2520. Its standard form is $2^3 \cdot 3^2 \cdot 5^1 \cdot 7^1$. By expanding the standard form, we get $2 \cdot 2 \cdot 2 \cdot 3 \cdot 3 \cdot 5 \cdot 7$ which is a prime factorization of 2520.

8   Consider the integer 51975. Its prime factorization is $3 \cdot 3 \cdot 3 \cdot 5 \cdot 5 \cdot 7 \cdot 11$. By collecting repetitions of primes as prime powers, we get $3^3 \cdot 5^2 \cdot 7^1 \cdot 11^1$ as the standard form of 51975.

9   Since the standard form and prime factorizations are so easily converted into each other, the terms "standard form" and "prime factorization" are often used interchangeably.

2.5 Finding the prime factorization of an integer

1   We have proven that each integer greater than 1 is the product of one or more prime numbers. In this section, we discuss a simple method of writing an integer greater than 1 as a product of one or more primes.

2   Let us find the prime factorization of 819. The primes not greater than $\sqrt{819}$ are 2,3,5,7,11,13,17,19,23.

3   Let $n_1 = 819$. We need to find a prime factor p_1 of n_1. Let us try to divide 819 by the primes $2, \ldots, 23$ in that order. We find that $3 \mid 819$. Hence $p_1 = 3$.

4   Let $n_2 = \frac{n_1}{p_1}$. Thus $n_2 = \frac{819}{3} = 273$. Hence, we need to find a prime factor p_2 of 273. The primes not greater than $\sqrt{273}$ are 2,3,5,7,11,13. After two divisions, we find that $3 \mid 273$. Thus $p_2 = 3$.

5   Let $n_3 = \frac{n_2}{p_2}$. Thus $n_3 = \frac{273}{3} = 91$. Hence, we need to find a prime factor p_3 of 91. The primes not greater than $\sqrt{91}$ are 2,3,5,7. After four divisions, we find that $7 \mid 91$. Thus $p_3 = 7$.

6   Let $n_4 = \frac{n_3}{p_3}$. Thus $n_4 = \frac{91}{7} = 13$. Hence, we need to find a prime factor p_4 of 13. The primes not greater than $\sqrt{13}$ are 2,3. After two divisions, we find that 13 is prime. Hence $p_4 = 13$.

7   Since $n_4 = 13$ is prime, we stop the process. Then the product $p_1 \cdot p_2 \cdot p_3 \cdot p_4$ is a prime factorization of n_1. Hence $3 \cdot 3 \cdot 7 \cdot 13$ is a prime factorization of 819.

Exercises 2.5

Find the prime factorization of each integer below.

1 104

2 1365

3 26411

4 2100875

2.6 GCD and LCM using prime factorization

1. In this section, we show how to use the standard form of an integer to find its GCD and LCM. Then we formalize this method and prove that it always works.

2. **GCD using prime factorization.** We can find the gcd of two integers greater than 1 if we are given their standard forms.

3. For example consider the integers 51975 and 2520. Their standard forms are given below.

$$51975 = 3^3 \cdot 5^2 \cdot 7^1 \cdot 11^1$$
$$2520 = 2^3 \cdot 3^2 \cdot 5^1 \cdot 7^1$$

4. To find the gcd of the integers, we first write the factorizations in terms of the primes that divide either of the integers. The primes are 2, 3, 5, 7, 11. Missing primes are represented as having zero exponent. Thus

$$51975 = 2^0 \cdot 3^3 \cdot 5^2 \cdot 7^1 \cdot 11^1$$
$$2520 = 2^3 \cdot 3^2 \cdot 5^1 \cdot 7^1 \cdot 11^0$$

5. We get the gcd by finding the minimum power of each prime in the expressions.

$$\begin{aligned}\gcd(a,b) &= 2^{\min(0,3)} \cdot 3^{\min(3,2)} \cdot 5^{\min(2,1)} \cdot 7^{\min(1,1)} \cdot 11^{\min(1,0)} \\ &= 2^0 \cdot 3^2 \cdot 5^1 \cdot 7^1 \cdot 11^0 \\ &= 1 \cdot 9 \cdot 5 \cdot 7 \cdot 1 \\ &= 315\end{aligned}$$

6 We prove this method always works and formally state it as a theorem below.

7 Let $p_1, \ldots, p_n$ be distinct prime numbers. Let $k_1, \ldots, k_n$ and $j_1, \ldots, j_n$ be nonnegative integers. Then,

$$\gcd(p_1^{k_1} \cdots p_n^{k_n}, p_1^{j_1} \cdots p_n^{j_n}) = p_1^{\min(k_1, j_1)} \cdots p_n^{\min(k_n, j_n)}.$$

Proof Let $c_1 = \min(k_1, j_1), \ldots, c_n = \min(k_n, j_n)$. Let $d = p_1^{c_1} \cdots p_n^{c_n}$, $a = p_1^{k_1} \cdots p_n^{k_n}$ and $b = p_1^{j_1} \cdots p_n^{j_n}$

Since $0 \leq c_i \leq k_i$ for all $1 \leq i \leq n$, it follows that

$$p_i^{c_i} \leq p_i^{k_i}$$
$$p_i^{c_i} \mid p_i^{k_i}$$

Since $p_i^{c_i} \mid p_i^{k_i}$ and $p_i^{k_i} \mid a$, it follows that $p_i^{c_i} \mid a$. Thus

$$p_1^{c_1} \mid a, \ldots, p_n^{c_n} \mid a$$

Since $p_1, \ldots, p_n$ are distinct prime numbers, it follows that $p_1^{c_1}, \ldots, p_n^{c_n}$ are pairwise coprime. Hence,

$$p_1^{c_1} \cdots p_n^{c_n} \mid a$$
$$d \mid a \tag{1}$$

Similarly,
$$d \mid b \tag{2}$$

From (1) and (2), we can conclude that d is a common divisor of a and b. Let m be a common divisor of a and b. Thus,

$$m \mid a$$
$$m \mid p_1^{k_1} \cdots p_n^{k_n}$$
$$m = p_1^{l_1} \cdots p_n^{l_n}$$

for some nonnegative integers $l_1, \ldots, l_n$. Let $1 \leq i \leq n$. Thus,

$$p_i^{l_i} \mid m.$$

Since $m \mid a$ and $m \mid b$, it follows that $p_i^{l_i} \mid a$ and $p_i^{l_i} \mid b$. Thus

$$p_i^{l_i} \mid p_1^{k_1} \cdots p_i^{k_i} \cdots p_n^{k_n}$$

$$p_i^{l_i} \mid p_i^{k_i} \cdot \frac{a}{p_i^{k_i}}$$

Since the integers $p_1^{k_1}, \ldots, p_n^{k_n}$ are pairwise coprime, $\gcd(p_i^{l_i}, \frac{a}{p_i^{k_i}}) = 1$. Thus,

$$p_i^{l_i} \mid p_i^{k_i}$$
$$p_i^{l_i} \leq p_i^{k_i}$$

Therefore, $l_i \leq k_i$. Similarly, we get $l_i \leq j_i$. Thus, $l_i \leq k_i$ and $l_i \leq j_i$. Hence $l_i \leq \min(k_i, j_i)$. Thus,

$$l_i \leq c_i$$
$$p_i^{l_i} \leq p_i^{c_i}$$

Thus,

$$p_1^{l_1} \leq p_1^{c_1}$$
$$p_2^{l_2} \leq p_2^{c_2}$$
$$\vdots$$
$$p_n^{l_n} \leq p_n^{c_n}$$

Multiply all the inequalities

$$p_1^{l_1} \cdots p_n^{l_n} \leq p_1^{c_1} \cdots p_n^{c_n}$$
$$m \leq d$$

Thus each common divisor of a, b is not greater than d. It follows that $d = \gcd(a, b)$. Hence

$$p_1^{c_1} \cdots p_n^{c_1} = \gcd(p_1^{k_1} \cdots p_n^{k_n}, p_1^{j_1} \cdots p_n^{j_n})$$
$$p_1^{\min(k_1, j_1)} \cdots p_n^{\min(k_n, j_n)} = \gcd(p_1^{k_1} \cdots p_n^{k_n}, p_1^{j_1} \cdots p_n^{j_n})$$

∎

8 **LCM using prime factorization.** We can use the standard forms of two integers to find their LCM. The process is similar to that of GCD. But instead of the min function we use the max function.

9 For example, consider the integers 51975 and 2520. Their standard forms are given below

$$51975 = 3^3 \cdot 5^2 \cdot 7^1 \cdot 11^1$$
$$2520 = 2^3 \cdot 3^2 \cdot 5^1 \cdot 7^1$$

10 The primes that divide at least one of the integers are 2,3,5,7,11. We can rewrite the standard forms to use these primes by allowing zero exponents:

$$51975 = 2^0 \cdot 3^3 \cdot 5^2 \cdot 7^1 \cdot 11^1$$
$$2520 = 2^3 \cdot 3^2 \cdot 5^1 \cdot 7^1 \cdot 11^0$$

11 Then the LCM is gotten thus,

$$\operatorname{lcm}(51975, 2520)$$
$$= 2^{\max(0,3)} \cdot 3^{\max(3,2)} \cdot 5^{\max(2,1)} \cdot 7^{\max(1,1)} \cdot 11^{\max(1,0)}$$
$$= 2^3 \cdot 3^3 \cdot 5^2 \cdot 7^1 \cdot 11^1$$
$$= 8 \cdot 27 \cdot 25 \cdot 7 \cdot 11$$
$$= 415800$$

12 We prove this method always works by formally stating it as a theorem below.

13 Let $p_1, \ldots, p_n$ be distinct prime numbers. Let $k_1, \ldots, k_n$ and $j_1, \ldots, j_n$ be non-negative integers. Then,

$$\operatorname{lcm}(p_1^{k_1} \cdots p_n^{k_n}, p_1^{j_1} \cdots p_n^{j_n}) = p_1^{\max(k_1, j_1)} \cdots p_n^{\max(k_n, j_n)}.$$

Proof I Let $c_1 = \max(k_1, j_1), \ldots, c_n = \max(k_n, j_n)$.

Thus for all $1 \le i \le n$, $c_i \ge k_i$ and $c_i \ge j_i$. Hence $k_i \le c_i$ and $j_i \le c_i$. Thus $p_i^{k_i} \le p^{c_i}$ and $p_i^{j_i} \le p^{c_i}$. Hence $p_i^{k_i} \le p^{c_i}$ and $p_i^{j_i} \le p^{c_i}$. Thus,

$$p_1^{k_1} \mid p_1^{c_1}$$
$$\vdots$$
$$p_n^{k_n} \mid p_n^{c_n}$$

Multiply all the statements together

$$p_1^{k_1} \cdots p_n^{k_n} \mid p_1^{c_1} \cdots p_n^{c_n}$$
$$a \mid d$$

Similarly, $b \mid d$. Since $a \mid d$ and $b \mid d$, we know that a, b are common multiples of d. Let m be a common multiple of a and b. Thus, for some integers j, k, $m = a \cdot k$ and $m = b \cdot j$. Hence, $m = p_1^{k_1} \cdots p_n^{k_n} \cdot k$ and $m = p_1^{j_1} \cdots p_n^{j_n} \cdot k$. Hence, $p_1^{k_1} \ldots p_n^{k_n}$ and $p_1^{j_1} \ldots p_n^{j_n}$ are all divisors of m. For all $1 \le i \le n$, $p_i^{k_i}$ and $p_i^{j_i}$ are divisors of m. Now, $\max(k_i, j_i) = k_i$ or $\max(k_i, j_i) = j_i$. In either case $p_i^{\max(k_i, j_i)}$ is a divisor of m. Thus $p_i^{c_i}$ is a divisor of m. Hence, $p_1^{c_1}, \ldots, p_n^{c_n}$ are divisors of m. Since the integers $p_1^{c_1}, \ldots, p_n^{c_n}$ are pairwise coprime, it follows that their product $p_1^{c_1} \cdots p_n^{c_n}$ is a divisor of m. Thus d is a divisor of m. Hence $d \mid m$. Thus $d \leq m$. Hence d is the least common multiple of a and b. Thus

$$d = \text{lcm}(a, b)$$
$$p_1^{c_1} \cdots p_n^{c_n} = \text{lcm}(p_1^{k_1} \cdots p_n^{k_n}, p_1^{j_1} \cdots p_n^{j_n})$$
$$p_1^{\max(k_1, j_1)} \cdots p_n^{\max(k_n, j_n)} = \gcd(p_1^{k_1} \cdots p_n^{k_n}, p_1^{j_1} \cdots p_n^{j_n})$$

■

Proof II Let $a = p_1^{k_1} \cdots p_n^{k_n}$ and $b = p_1^{j_1} \cdots p_n^{j_n}$. Let $c_1 = \min(k_1, j_1), \ldots, c_n = \min(k_n, j_n)$. Now,

$$a \cdot b = p_1^{k_1} \cdots p_n^{k_n} \cdot p_1^{j_1} \cdots p_n^{j_n}$$
$$= p_1^{k_1+j_1} \cdots p_n^{k_n+j_n}$$

And,

$$\gcd(a, b) = p_1^{c_1} \cdots p_n^{c_n}$$

We know that

$$\text{lcm}(a, b) = \frac{a \cdot b}{\gcd(a, b)}$$
$$= \frac{p_1^{k_1+j_1} \cdots p_n^{k_n+j_n}}{p_1^{c_1} \cdots p_n^{c_n}}$$
$$= p_1^{k_1+j_1-c_1} \cdots p_n^{k_n+j_n-c_n}$$

Since $\min(x, y) = x + y - \max(x, y)$

$$= p_1^{\min(k_1, j_1)} \cdots p_n^{\min(k_n, j_n)}$$

■

Exercises 2.6

1 Find the GCD of the following integers

    a $5^3 \cdot 7^2$ and 5^7

    b $2^4 \cdot 5^{11}$ and $3^2 \cdot 5 \cdot 7^3$

2 Find the LCM of the following integers

  a $2^2 \cdot 3^4$ and $5^3 \cdot 7$

  b $2 \cdot 11 \cdot 17^2$ and $11^2 \cdot 17$

3 Find the GCD and LCM of each pair of integers below.

  a $2^5 \cdot 3 \cdot 7^6$ and $3^2 \cdot 7 \cdot 19$

  b $13 \cdot 11^2$ and $2 \cdot 7^3 \cdot 13$

4 Let $p_1, \ldots, p_n$ be distinct prime numbers. Let $j_1, \ldots, j_n, k_1, \ldots, k_n$ and $l_1, \ldots, l_n$ be nonnegative integers. Prove that

  a $\gcd(p_1^{k_1} \cdots p_n^{k_n}, p_1^{j_1} \cdots p_n^{j_n}, p_1^{l_1} \cdots p_n^{l_n}) = p_1^{\min(k_1, j_1, l_1)} \cdots p_n^{\min(k_n, j_n, l_n)}$

  b $\text{lcm}(p_1^{k_1} \cdots p_n^{k_n}, p_1^{j_1} \cdots p_n^{j_n}, p_1^{l_1} \cdots p_n^{l_n}) = p_1^{\max(k_1, j_1, l_1)} \cdots p_n^{\max(k_n, j_n, l_n)}$

5 Find the GCD and LCM of each triple of integers below.

  a $7^2 \cdot 13$, $2^3 \cdot 3^2$ and $13 \cdot 11^2$

  b $7 \cdot 19$, $3^2 \cdot 7 \cdot 19$ and $2 \cdot 7^3 \cdot 13$

CHAPTER

3

Infinitude of primes

3.1 Introduction

1 Euclid proved that there are infinitely many primes. In this section, we provide Euclid's proof of this theorem. Then we provide another proof that involves factorials.

2 **Euclid's proof.** Here we describe Euclid's proof that there are infinitely many primes.

3 There are infinitely many primes

Proof To prove that there are infinitely many primes, we only need to show that every finite list of primes is incomplete. Let $p_1, \ldots, p_n$ be prime numbers where n is a positive integer. Let $k = p_1 \cdots p_n$. Since each prime in $p_1, \ldots, p_n$ divides k, it follows that each prime in the list does not divide $k + 1$. Since $k + 1 > 1$, it follows that $k + 1$ has a prime divisor q. Thus q is not in the list. It follows that the list is incomplete. Therefore, there are infinitely many primes. ∎

4 **Factorial proof.** Here, we prove that there are infinitely many primes using factorials. Before this, we prove some lemmas.

5 For all positive integers n, all factors of $n! + 1$ greater than 1 are greater than n.

Proof Let n be a positive integer. Let d be a factor of $n! + 1$ such that $d > 1$. Since

$$n! \geq 1$$
$$n! + 1 \geq 2$$

Thus $n! + 1 > 1$. Since $d \mid n! + 1$ and $d > 1$, it follows that

$$d \nmid n!$$
$$d \nmid 1 \cdots n$$

Hence d is not in the list $1, \ldots, n$. Since $d > 1$ and d is not in $1, \ldots, n$, it follows that $d > n$. Thus all factors of $n! + 1$ that are greater than 1 are also greater than n ∎

6 For each positive integer n, there exists a prime number greater than n.

Proof Let n be a positive integer. Thus $n! + 1$ is an integer greater than 1. Hence, $n! + 1$ has a prime factor p. Since $p > 1$ and $p \mid n! + 1$, by a previous theorem, $p > n$. Hence there exists a prime number greater than n. ∎

7 We can use the above theorem to show that there are infinitely many primes

8 There are infinitely many primes.

Proof To prove that there are infinitely many primes, we simply show that there is no greatest prime. Let p be a prime number. Since p is an integer, by a previous theorem, there exists some prime number greater than p. Thus p is not the greatest prime. Hence there is no greatest prime. Therefore there are infinitely many primes. ∎

3.2 Sequence of pairwise coprime integers

1 In this section, we define sequence of pairwise coprime integers and provide several examples of such sequences. Then we show that such sequences can be used to prove the infinitude of primes. We also discuss how to use such sequences as an upper bound for the prime numbers.

2 **Definition.** Let $a_1, a_2, \ldots$ be a sequence of integers greater than 1. If every two integers that are at different positions in the sequence are coprime, we say the sequence is a sequence of pairwise coprime integers greater than 1.

3 **Example 1.** Every two different primes are coprime. Thus the sequence of primes is a sequence of pairwise coprime integers greater than 1:

$$2, 3, 5, 7, 11, \ldots$$

Example 2. Another example is the sequence $a_1, a_2, \ldots$ which is defined thus

$$a_1 = 2$$
$$a_n = a_1 \cdots a_{n-1} + 1 \quad \text{if } n > 1$$

Thus

$$a_1 = 2$$
$$a_2 = 2 + 1 = 3$$
$$a_3 = 2 \cdot 3 + 1 = 7$$
$$a_4 = 2 \cdot 3 \cdot 7 + 1 = 43$$

The first few terms of the sequence are 2, 3, 7, 43, 1807. Below, we show that this sequence is a sequence of pairwise coprime integers.

For all positive integers n, m, if $n \neq m$ then $\gcd(a_n, a_m) = 1$.

Proof Without loss of generality, let $m < n$. Hence $n > m$ and $m \geq 1$. Thus $n > 1$. Hence, by the definition of the sequence,

$$a_n = a_1 \cdots a_{n-1} + 1 \tag{1}$$

Since $0 < m < n$, it follows that m is in the list $1, \ldots, n - 1$. Thus a_m is in the list $a_1, \ldots, a_{n-1}$. Hence

$$a_m \mid a_1 \cdots a_{n-1}$$

It follows that

$$a_1 \cdots a_{n-1} = a_m \cdot k \tag{2}$$

for some integer k. We know that

$$\gcd(a_m \cdot k, a_m \cdot k + 1) = 1$$

Thus $\gcd(a_m \cdot k, a_n) = 1$. Hence $\gcd(a_m, a_n) = 1$ ∎

Example 3. We can create other sequences of pairwise coprime integers by changing the initial value of a_1 to some value greater than 2. For example, if $a_1 = 3$, we get

$$a_1 = 3$$
$$a_2 = 3 + 1 = 4$$
$$a_3 = 3 \cdot 4 + 1 = 13$$
$$a_4 = 3 \cdot 4 \cdot 13 + 1 = 157$$

Chapter 3 Infinitude of primes

9 The above sequence is also a sequence of pairwise coprime integers. In general, due to the previous theorem, all sequences created with the rule $a_n = a_1 \cdots a_{n-1} + 1$ are sequences of pairwise coprime integers.

10 **More examples.** There are other rules for constructing sequences where all terms are pairwise coprime. Consider the following sequence

$$a_1 = 3$$
$$a_n = a_1 \cdots a_{n-1} - 1 \quad \text{if } n > 1$$

Thus,

$$a_1 = 3$$
$$a_2 = 3 - 1 = 2$$
$$a_3 = 3 \cdot 2 - 1 = 5$$
$$a_4 = 3 \cdot 2 \cdot 5 - 1 = 29$$

11 Other rules are given below

1. $a_n = a_{n-1}! + 1$
2. $a_n = a_{n-1}! - 1$

12 **Infinitude of primes.** There is another proof of the infinitude of primes. This proof utilizes a sequence of pairwise coprime integers. It shows that any pair of integers in the sequence is relatively prime. Thus no two positive integers in the list will have the same prime divisor. Since the list is infinite then the list of primes must also be infinite. Such a proof is given below.

13 Let $a_1, a_2, \ldots$ be a sequence of pairwise coprime integers greater than 1. Then there are infinitely many primes.

Proof For all positive integers n, let q_n be the least prime factor of a_n. Let n, m be two distinct primes of the sequence $a_1, a_2, \ldots$. Since $\gcd(n, m) = 1$, it follows that n and m have no common prime factor. Hence the least prime factor of n is different from the least prime factor of m. Thus the sequence $q_1, q_2, \ldots$ is a sequence of distinct primes. Since the sequence $a_1, a_2, \ldots$ is infinite, it follows that the sequence $q_1, q_2, \ldots$ is infinite. Therefore there are infinitely many primes. ∎

14 **Upper bound for prime numbers.** Consider the sequence $a_1, a_2, \ldots$ defined thus:

$$a_1 = 2$$

$$a_n = a_1 \cdots a_{n-1} + 1 \quad \text{if } n > 1$$

Then the sequence is $2, 3, 7, \ldots$. From an earlier example, we know that this is a sequence of pairwise coprime integers. Let us compare this sequence with the sequence of prime numbers $p_1, p_2, \ldots$. Observe that

$$2 \leq 2, 3 \leq 3, 5 \leq 7$$

Thus

$$p_1 \leq a_1, \ p_2 \leq a_2, \ p_3 \leq a_3$$

15 In general $p_n \leq a_n$ for all positive integers n. This is also true regardless of the sequence of pairwise coprime integers we use for $a_1, a_2, \ldots$, provided that the sequence is increasing, and all terms are greater than 1. We prove this below.

16 Let $a_1, a_2, \ldots$ be a sequence of pairwise coprime integers greater than 1 where $a_1 \leq a_2 \leq \cdots$. Let p_n be the nth prime number for each positive integer n. Then $p_n \leq a_n$ for all positive integers n.

Proof For each positive integer n, let q_n be the least prime factor of a_n. Since $a_1, a_2, \ldots$ are pairwise coprime, it follows that $q_1, q_2, \ldots$ are distinct prime numbers.

Let n be a positive integer. There are exactly $n - 1$ prime numbers less than p_n. The list $q_1, \ldots, q_n$ contains exactly n distinct primes. Hence it is not the case that all the primes in $q_1, \ldots, q_n$ are less than p_n. Thus some prime in $q_1, \ldots, q_n$ is not less than p_n. Hence, for some integer i where $1 \leq i \leq n$,

$$p_n \leq q_i$$

Since q_i is a prime factor of a_i, it follows that

$$q_i \leq a_i$$

Since $a_n \geq a_{n-1} \geq \cdots \geq a_1$, it follows that $a_n \geq a_i$. Thus $a_n \geq a_i$ and $a_i \geq q_i$. Hence, $a_n \geq q_i$. Thus $a_n \geq q_i$ and $q_i \geq p_n$. Therefore $a_n \geq p_n$. ∎

3.3 Inequalities of prime numbers

1 In this section, we prove the following inequalities involving the sequence of prime numbers:

- $p_n < p_1 \cdots p_{n-1}$ if $n > 2$
- $p_{n+1} < p_n^n$ if $n > 1$

2 Let p_n be the nth prime number for each positive integer n

3 $p_n \leq p_1 \cdots p_{n-1} + 1$ for $n > 1$

Proof Let n be an integer greater than 1. Let

$$k = p_1 \cdots p_{n-1} + 1$$

Hence $k > 1$. Thus k has a prime divisor, p. Hence $p \mid k$. Thus $p \nmid k - 1$. Hence $p \nmid p_1 \cdots p_{n-1}$. Thus p is not in the list $p_1, \ldots, p_{n-1}$. Thus $p > p_{n-1}$. Hence,

$$p \geq p_n \tag{1}$$

Since $p \mid k$, it follows that

$$p \leq k$$
$$p \leq p_1 \cdots p_{n-1} + 1 \tag{2}$$

From (1) and (2) we get $p_n \leq p$ and $p \leq p_1 \cdots p_{n-1} + 1$. Thus

$$p_n \leq p_1 \cdots p_{n-1} + 1$$

∎

4 For all integers n, if $n > 2$ then

$$p_n \leq p_1 \cdots p_{n-1} - 1$$

Proof Let n be an integer greater than 2. Let $k = p_1 \cdots p_{n-1} - 1$. Since $n > 2$, it follows that $n - 1 > 1$. Thus $n - 1 \geq 2$. Hence $p_{n-1} \geq p_2$. Thus

$$p_{n-1} \geq 3$$
$$p_1 \cdots p_{n-1} \geq 3$$
$$p_1 \cdots p_{n-1} - 1 \geq 2$$
$$k \geq 2$$
$$k > 1$$

Since $k > 1$, it follows that k has a prime divisor p. Hence $p \mid k$. Thus $p \nmid k+1$. Hence, $p \nmid p_1 \cdots p_{n-1}$. Thus p is not in the list $p_1, \ldots, p_{n-1}$. Hence $p > p_{n-1}$. It follows that
$$p \geq p_n \tag{1}$$
Since $p \mid k$, it follows that
$$p \leq k$$
$$p \leq p_1 \cdots p_{n-1} - 1 \tag{2}$$
From (1) and (2) we get $p_n \leq p$ and $p \leq p_1 \cdots p_{n-1} - 1$. Thus
$$p_n \leq p_1 \cdots p_{n-1} - 1$$

∎

5 Let n be an integer greater than 2. Then
$$p_n < p_1 \cdots p_{n-1}$$

Proof Let n be an integer greater than 2. Thus
$$p_n \leq p_1 \cdots p_{n-1} - 1$$
$$p_n < p_1 \cdots p_{n-1}$$

∎

6 Let n be an integer greater than 1. Then
$$p_{n+1} < p_n^n$$

Proof Let n be an integer greater than 1. Hence $n + 1 > 2$. By a previous theorem,
$$p_{n+1} < p_1 \cdots p_n \tag{1}$$
We know that
$$p_1 \leq p_n$$
$$p_2 \leq p_n$$
$$\vdots$$
$$p_n \leq p_n$$
Multiply all the inequalities together
$$p_1 \cdots p_n \leq p_n^n \tag{2}$$
From (1) & (2) we get
$$p_{n+1} < p_n^n$$

∎

Exercises 3.3

1 Prove that for each integer n greater than 1, there exists a prime p such that $p_n < p < p_n^n$.

2 Let $a_1, a_2, \ldots$ be pairwise coprime integers greater than 1. Prove that there are infinitely many primes by using the sequence $a_1, a_2, \ldots$ and the sequence $q_1, q_2, \ldots$ where q_n is the greatest prime factor of a_n for each positive integer n.

3 Let $a_1, a_2, \ldots$ be a sequence of integers greater than 1 where the terms are pairwise coprime.

  a For each positive integer n, let $b_n = a_n^2$. Prove that the sequence $b_1, b_2, \ldots$ is a sequence of pairwise coprime integers greater than 1.

  b For each positive integer n, let $b_n = a_n^3$. Prove that the sequence $b_1, b_2, \ldots$ is a sequence of pairwise coprime integers greater than 1.

4 Prove that there are infinitely many composite integers

5 Let $a_1, a_2, \ldots$ be a sequence of integers defined as given below. Show that the terms are pairwise coprime.

  a
  $$a_1 = 2$$
  $$a_n = a_{n-1}! + 1 \quad \text{if } n > 1$$

  b
  $$a_1 = 3$$
  $$a_n = a_{n-1}! - 1 \quad \text{if } n > 1$$

CHAPTER 4
Forms of primes

4.1 Mersenne numbers

1. In this section, we define Mersenne numbers and their notation. Then we prove a property of primes that are one less than a power.

2. **Definition.** A Mersenne number is an integer of the form

$$2^n - 1$$

where n is a positive integer.

3. A Mersenne prime is a Mersenne number that is prime. Examples of such primes are given below.

$$2^2 - 1 = 3$$
$$2^3 - 1 = 7$$
$$2^5 - 1 = 31$$
$$2^7 - 1 = 127$$
$$2^{13} - 1 = 8191$$
$$2^{17} - 1 = 131071$$

4. Observe that the exponents are $2, 3, 5, 7, 13, 17$. They are all primes. In this section, we prove that the exponent of a Mersenne prime is always prime. Moreover, we show that if a prime is of the form $a^p - 1$, where $a, p > 1$ then $a = 2$ and p is prime.

5. As of 2024, the largest known prime number is a Mersenne prime:

$$2^{136279841} - 1.$$

It is conjectured that there are infinitely many Mersenne primes.

6 **Notation for Mersenne numbers.** For each positive integer n, we use M_n to denote the nth Mersenne number:

$$M_n = 2^n - 1$$

7 **Primes that are one less than a power.** In this section we prove that if $a^p - 1$ is prime then $a = 2$ and p is prime. Then we use this theorem to show that 7 is the only prime that is one less than a cube.

8 Let a, p be integers greater than 1. If $a^p - 1$ is prime then $a = 2$.

Proof From a geometric series,

$$a^p - 1 = (a - 1)(a^0 + a^1 + \cdots + a^{p-1}).$$

Thus $a - 1$ is a divisor of $a^p - 1$. Since $a > 1$, it follows that $a - 1 > 0$. Thus $a - 1$ is a positive divisor of $a^p - 1$. Since $a^p - 1$ is prime, its positive divisors are limited to 1 and itself. Thus,

$$a - 1 = 1 \quad \text{or} \quad a - 1 = a^p - 1$$

$$a = 2 \quad \text{or} \quad a^1 = a^p$$

$$a = 2 \quad \text{or} \quad p = 1$$

Since $p > 1$, it follows that $p \neq 1$. Hence $a = 2$. ∎

9 Let p be an integer greater than 1. If $2^p - 1$ is a prime then p is prime.

Proof To prove that p is prime, we simply show that each of its positive divisors is 1 or p. Let r be a positive divisor of p. Hence $p = rs$ for some positive integers r, s. Thus

$$2^p - 1 = 2^{rs} - 1$$
$$= (2^r)^s - 1$$

From a geometric series,

$$= (2^r - 1)((2^r)^0 + \cdots + (2^r)^{s-1}) \tag{1}$$

Since $r \geq 1$, it follows that $2^r \geq 2^1$. Hence $2^r \geq 2$. Thus $2^r - 1 \geq 1$. Hence, from (1), we know that $2^r - 1$ is a positive divisor of $2^p - 1$. Since $2^p - 1$ is prime, its positive divisors are 1 and itself. Hence

$$2^r - 1 = 1 \quad \text{or} \quad 2^r - 1 = 2^p - 1$$

$$2^r = 2^1 \quad \text{or} \quad 2^r = 2^p$$

$$r = 1 \quad \text{or} \quad r = p$$

Hence each positive divisor of p is either 1 or p. It follows that p is prime. ∎

10 Let a, p be integers greater than 1. If $a^p - 1$ is prime then $a = 2$ and p is prime.

Proof Since $a^p - 1$ is prime, it follows that $a = 2$. Thus $a^p - 1 = 2^p - 1$. Hence $2^p - 1$ is prime. It follows that p is prime. Thus $a = 2$ and p is prime. ∎

11 Using the above theorem we can prove that the only prime of the form $n^3 - 1$ is 7.

12 The only prime of the form $n^3 - 1$ is 7.

Proof Let n be an integer such that $n^3 - 1$ is prime. It follows that $n^3 - 1 > 0$. Thus $n^3 > 1$. Hence $n > \sqrt[3]{1}$. Thus $n > 1$. Since $n > 1$ and $n^3 - 1$ is prime, it follows that $n = 2$. Thus $n^3 - 1 = 2^3 - 1 = 7$. Hence the only prime of the form $n^3 - 1$ is 7. ∎

4.2 Sums and products

1 In this section, we prove the following results: Let r be a real number such that $-1, 0 \neq r$. Then for all positive integers n,

- if n is odd then $r^n + 1 = (r+1)(r^0 - r^1 + \cdots + r^{n-1})$
- if n is even then $r^n - 1 = (r+1)(r^{n-1} - \cdots + r^1 - r^0)$
- $(2^{2^0} + 1)(2^{2^1} + 1) \cdots (2^{2^{n-1}} + 1) = 2^{2^n} - 1$

These results will help us understand Fermat numbers. Before this, we prove the following theorem.

2 Let r be a real number such that $-1, 0 \neq r$. Then for all positive integers n,

$$r^0 - r^1 + \cdots - r^{n-1}(-1)^n = \frac{1 - r^n(-1)^n}{1 + r}$$

Proof Since $r \neq -1$, $r + 1 \neq 0$. For each nonnegative integer n, let

$$a_n = \frac{r^n}{r+1}$$

For each positive integer n, let

$$b_n = a_n + a_{n-1}$$

Thus,

$$b_n = \frac{r^n}{r+1} + \frac{r^{n-1}}{r+1}$$
$$= r^{n-1}$$

Since $b_n = a_n + a_{n-1}$, by the property of alternating sums,

$$b_1 - b_2 + \cdots - b_n(-1)^n = a_0 - a_n(-1)^n$$

Thus,

$$r^0 - r^1 + \cdots - r^{n-1}(-1)^n$$
$$= \frac{r^0}{r+1} - \frac{r^n}{r+1}(-1)^n$$
$$= \frac{1}{r+1} - \frac{r^n}{r+1}(-1)^n$$
$$= \frac{1 - r^n(-1)^n}{r+1}$$

∎

3 Let r be a real number such that $r \neq -1$ and $r \neq 0$. Then for all positive integers n, if n is odd.

$$r^n + 1 = (r+1)(r^0 - r^1 + \cdots + r^{n-1})$$

Proof From a previous theorem,

$$r^0 - r^1 + \cdots - r^{n-1}(-1)^n = \frac{1 - r^n(-1)^n}{r+1}$$

Since n is odd, $(-1)^n = -1$. Thus,

$$r^0 - r^1 + \cdots - r^{n-1}(-1) = \frac{1 - r^n(-1)}{r+1}$$

$$r^0 - r^1 + \cdots + r^{n-1} = \frac{1 + r^n}{r+1}$$

Multiply both sides by $r+1$

$$(r+1)(r^0 - r^1 + \cdots + r^{n-1}) = r^n + 1$$

∎

4. Let r be a real number such that $-1, 0 \neq r$. Then for all positive integers n, if n is even then

$$r^n - 1 = (r+1)(r^{n-1} - \cdots + r^1 - r^0)$$

Proof From a previous theorem

$$r^0 - r^1 + \cdots - r^{n-1}(-1)^n = \frac{1 - r^n(-1)^n}{r+1}$$

Since n is even, $(-1)^n = 1$. Thus,

$$r^0 - r^1 + \cdots - r^{n-1}(1) = \frac{1 - r^n(1)}{r+1}$$

Multiply both sides by -1

$$-r^0 + r^1 - \cdots + r^{n-1} = \frac{r^n - 1}{r+1}$$

Multiply both sides by $r+1$

$$(r+1)(r^{n-1} - \cdots + r^1 - r^0) = r^{n-1}$$

∎

5. For all positive integers n,

$$(2^{2^0} + 1)(2^{2^1} + 1) \cdots (2^{2^{n-1}} + 1) = 2^{2^n} - 1$$

Proof For each nonnegative integer n, let $a_n = 2^{2^n} - 1$. For each positive integer n, let

$$b_n = \frac{a_n}{a_{n-1}}$$
$$= \frac{2^{2^n} - 1}{2^{2^{n-1}} - 1}$$
$$= 2^{2^{n-1}} + 1$$

By the telescoping property

$$b_1 \cdot b_2 \cdots b_n = \frac{a_n}{a_0}$$

Thus

$$(2^{2^0} + 1)(2^{2^1} + 1) \cdots (2^{2^{n-1}} + 1) = \frac{a_n}{a_0}$$
$$= \frac{2^{2^n} - 1}{2^{2^0} - 1}$$
$$= 2^{2^n} - 1$$

∎

4.3 Fermat numbers

1 In this section, we define Fermat numbers and their notation. We prove some basic properties of Fermat numbers and show that these numbers are pairwise coprime. This property is used to show that there are infinitely many primes and establish an upper bound for the prime numbers.

Then we prove that if a prime is one more than a power then the base of the power is even and the exponent is a power of two.

2 **Definition of Fermat number.** A Fermat number is an integer of the form $2^{2^n} + 1$ where n is a nonnegative integer.

3 A Fermat prime is a Fermat number that is prime. Examples of such primes are given below.

$$2^{2^0} + 1 = 3$$
$$2^{2^1} + 1 = 5$$

$$2^{2^2} + 1 = 17$$
$$2^{2^3} + 1 = 257$$
$$2^{2^4} + 1 = 65537$$

4   These are currently the only known Fermat primes. It is conjectured that there are no other such primes.

5   **Notation for Fermat numbers.** For each nonnegative integer n, we use F_n to denote the nth Fermat number:

$$F_n = 2^{2^n} + 1.$$

6   **Basic properties of Fermat numbers.** Here, we prove some basic properties of Fermat numbers:

- $F_n > F_{n-1}$ for all positive integers n
- $F_0 < F_1 < F_2 < \cdots$
- $F_n > 1$ for all nonnegative integers n

7   $F_n > F_{n-1}$ for all positive integers n

Proof   Let n be a positive integer. Since $2^n > 2^{n-1}$, it follows that $2^{2^n} > 2^{2^{n-1}}$. Adding 1 to both sides of the inequality, we get $2^{2^n} + 1 > 2^{2^{n-1}} + 1$. Hence $F_n > F_{n-1}$. ∎

8   $F_0 < F_1 < F_2 < \cdots$

Proof   For all positive integers n, $F_{n-1} < F_n$. Hence,

$$F_0 < F_1$$
$$F_1 < F_2$$
$$F_2 < F_3$$
$$\vdots$$

Therefore, $F_0 < F_1 < F_2 < \cdots$. ∎

9   $F_n > 1$ for all nonnegative integers n

Proof Since $F_0 < F_1 < F_2 < \ldots$, it follows that the sequence of Fermat numbers is an increasing sequence. The least Fermat number F_0 is greater than 1. Hence all Fermat numbers are greater than 1. ∎

10 **Fermat numbers are pairwise coprime.** Here, we show that Fermat numbers are pairwise coprime. This property is used to prove that there are infinitely many primes and

$$p_n \leq F_{n-1} \quad \text{if } n > 0.$$

Before this, we prove the following lemma.

11 For all nonnegative integers m, n, if $n > m$ then

$$F_m \mid F_n - 2$$

Proof I In this proof, we prove the statement by using a telescoping product. Let m, n be nonnegative integers such that $n > m$. Hence $n > m$ and $m \geq 0$. It follows that $n > 0$. Since n is a positive integer,

$$(2^{2^0} + 1)(2^{2^1} + 1) \cdots (2^{2^{n-1}} + 1) = 2^{2^n} - 1$$

Since $0 \leq m \leq n - 1$, it follows that m is among the integers $0, 1, \ldots, n-1$. Thus we may write

$$(2^{2^0} + 1)(2^{2^1} + 1) \cdots (2^{2^m} + 1) \cdots (2^{2^{n-1}} + 1) = 2^{2^n} - 1$$

Thus

$$F_0 \cdot F_1 \cdots F_m \cdots F_{n-1} = 2^{2^n} - 1$$

Hence

$$F_m \mid 2^{2^n} - 1$$

Since $F_n - 2 = 2^{2^n} - 1$, we may write

$$F_m \mid F_n - 2$$

∎

4.3 Fermat numbers

Proof II In this proof, we prove the statement by using an alternating sum. Let $k = n - m$. Thus $k > 0$. Hence 2^k is an even integer. Let $r = 2^{2^m}$. Thus r is a positive integer. Hence $r + 1 \mid r^{2^k} - 1$. Now,

$$r + 1 = 2^{2^m} + 1$$
$$= F_m$$

And,

$$r^{2^k} - 1 = (2^{2^m})^{2^k} - 1$$
$$= 2^{2^m \cdot 2^k} - 1$$
$$= 2^{2^{m+k}} - 1$$
$$= 2^{2^n} - 1$$
$$= F_n - 2$$

Thus,

$$F_m \mid F_n - 2$$

∎

12 For all nonnegative integers n, m, if $n \neq m$ then

$$\gcd(F_n, F_m) = 1.$$

Proof Without loss of generality, let $n \geq m$. Since $n \neq m$, it follows that $n > m$. Thus

$$F_m \mid F_n - 2$$

Let d be a positive integer such that $d \mid F_m$ and $d \mid F_n$. Since $d \mid F_m$ and $F_m \mid F_n - 2$, it follows that $d \mid F_n - 2$. Hence $d \mid F_n$ and $d \mid F_n - 2$. Thus d divides all linear combinations of F_n and $F_n - 2$.

Observe that

$$2 = F_n(1) + (F_n - 2)(-1)$$

Hence 2 is a linear combination of F_n and $F_n - 2$. Thus $d \mid 2$. Hence $d \leq 2$. Thus $1 \leq d \leq 2$. Hence $d = 1$ or $d = 2$. Since F_n is odd, $2 \nmid F_n$. Hence, $d \neq 2$. Thus $d = 1$. It follows that the only common divisor of F_n and F_m is 1. Hence

$$\gcd(F_n, F_m) = 1$$

∎

13 There are infinitely many primes

Proof Consider the following sequence of integers: $F_0, F_1, F_2, \ldots$. Hence each integer in the sequence is greater than 1. From a previous theorem, $\gcd(F_n, F_m) = 1$ for all distinct nonnegative integers n, m. Hence the integers in the sequence $F_0, F_1, \ldots$ are pairwise coprime integers greater than 1. Therefore there are infinitely many primes. ∎

14

Let p_n be the nth prime number for each positive integer n. Then for all positive integers n,
$$p_n \leq F_{n-1}.$$

Proof

For each nonnegative integer n, let $F_n = 2^{2^n} + 1$. Thus the sequence $F_0, F_1, \ldots$ consists of pairwise coprime integers greater than 1. Also $F_0 < F_1 < \cdots$. By a previous theorem, for each positive integer n,
$$p_n \leq F_{n-1}.$$

∎

15 **Primes that are one more than a power.**
Here we show that if $a^n + 1$ is prime then a is even and $n = 2^m$ for some nonnegative integer m. Before proving this theorem, we need the following lemma.

16
For all positive integers n, $n = 2^k$ for some nonnegative integer k if and only if 1 is the only odd positive factor of n.

Proof

Part 1: If $n = 2^k$ for some nonnegative integer k then 1 is the only odd positive factor of n Let n be a positive integer such that $n = 2^k$ for some nonnegative integer k. Let d be an odd positive divisor of n. Since d is a positive divisor of n,
$$\gcd(n, d) = d \qquad (1)$$

Since d is an odd integer,
$$\gcd(2, d) = 1$$

4.3 Fermat numbers

$$\gcd(2^k, d) = 1$$
$$\gcd(n, d) = 1 \quad (2)$$

From (1) and (2), we get
$$d = 1$$

Hence each odd positive divisor of n is 1. Thus 1 is the only odd positive divisor of n.

Part 2: If 1 is the only odd positive divisor of n then $n = 2^k$ for some nonnegative integer k Let n be a positive integer such that 1 is the only odd positive divisor of n. It is either $n = 1$ or $n > 1$. If $n = 1$ then $n = 2^0$. Thus $n = 2^k$ for some nonnegative integer k.

If $n > 1$ then n has a prime divisor p. Since $p \neq 1$ and $p \mid n$, it follows that p is not odd. Thus p is even. Since 2 is the only even prime, it follows that $p = 2$. Hence 2 is the only prime divisor of n. Due to the fundamental theorem of arithmetic,
$$n = 2 \cdots 2$$

Hence $n = 2^k$ for some nonnegative integer k.

Therefore, in both cases, $n = 2^k$ for some nonnegative integer k. ∎

17 Let a, n be integers greater than 1. If $a^n + 1$ is prime then a is even and $n = 2^m$ for some nonnegative integer m.

Proof Let a, n be integers greater than 1 such that $a^n + 1$ is prime. First, we show that a is even. Then we show that n is a power of 2.

The integer a is even Since $a, n \geq 2$, it follows that $a^n \geq 2^2$. Thus, $a^n > 3$. Hence $a^n + 1 > 4$. Thus $a^n + 1 > 2$. All prime numbers greater than 2 are odd. Thus $a^n + 1$ is odd. Hence a^n is even. It follows that a is even.

For some nonnegative integer m, $n = 2^m$ Assume n has an odd factor k which is greater than 1. Thus $n = kj$ for some positive integer j.

$$a^n + 1 = a^{kj} + 1$$
$$= (a^j)^k + 1$$

From an alternating sum,

$$= (a^j + 1)\left[(a^j)^0 - (a^j)^1 + \cdots - (a^j)^{k-1}\right]$$

Since $1 < a^j + 1 < a^n + 1$, and $a^j + 1 \mid a^n + 1$, it follows that $a^n + 1$ is composite. Thus, $a^n + 1$ is composite. Thus, $a^n + 1$ is prime and composite. This contradiction implies that n has no odd factors greater than 1. Hence, $n = 2^m$ for some nonnegative integer m.

Conclusion In conclusion, a is even and $n = 2^m$ for some nonnegative integer m. ∎

18 We can use the above theorem to prove the following corollary.

19 For all integers n greater than 1, $n^3 + 1$ is composite.

Proof Assume $n^3 + 1$ is prime. Thus 3 is a power of two. This is absurd. Hence $n^3 + 1$ is composite. ∎

Exercises 4.3

Prove the following.

1 Let n be a positive integer. Let p_n be the nth prime.

   a $p_n \leq 2^{2^{n-1}} + 1$

   b $p_n < 2^{2^n}$

2 For all positive integers n, $8^n + 1$ is composite.

(Hint: Use contradiction. Observe that $8^n + 1 = 2^{3n} + 1$.)

3 For all positive integers n, $216^n + 1$ is composite.

(Hint: Use contradiction. Observe that $216^n + 1 = 6^{3n} + 1$.)

4.4 Partition into primes

1 Consider the integer 10. We can write this integer as a sum of primes:

$$10 = 2 + 3 + 5$$

In this section, we show that each integer greater than 1 is either prime or a sum of primes. In order to prove this theorem, we first prove two lemmas:

- Each even integer greater than 2 is a sum of primes.
- Each odd integer greater than 3 is a sum of primes.

2 Each even integer greater than 2 is a sum of primes.

Proof Let n be an even integer greater than 2. Thus $n > 2$. Since n is even, $n = 2k$ for some integer k. Thus,

$$2k > 2$$
$$k > 1$$

Hence

$$2k = \underbrace{2 + \cdots + 2}_{k}$$
$$n = 2 + \cdots + 2$$

Since 2 is prime, it follows that $2 + \cdots + 2$ is a sum of primes. Hence n is a sum of primes. ∎

3 Each odd integer greater than 3 is a sum of primes.

Proof Let n be an odd integer greater than 3. Thus

$$n > 3 \qquad (1)$$

Since n is odd, $n = 2k + 3$ for some integer k. From (1), we know that

$$2k + 3 > 3$$
$$2k > 0$$
$$k > 0$$

Thus,

$$2k = \underbrace{2 + \cdots + 2}_{k}$$

$$2k + 3 = 2 + \cdots + 2 + 3$$

Since 2 and 3 are primes, it follows that $2 + \cdots + 2 + 3$ is a sum of primes. Hence n is a sum of primes. ∎

4   Each integer greater than 3 is a sum of primes.

Proof   Let n be an integer greater than 3. Thus n is even or n is odd. If n is even then n is an even integer greater than 3. Thus n is an even integer greater than 2. Hence n is a sum of primes.

If n is odd then n is an odd integer greater than 3. By a previous theorem, n is a sum of primes. Thus in either case n is a sum of primes. ∎

4.5   Goldbach conjecture

1   The Goldbach conjecture states that each even integer greater than 2 is the sum of two primes.

2   Examples of even integers and their representation as a sum of two primes are shown below.

$$4 = 2 + 2$$
$$6 = 3 + 3$$
$$8 = 3 + 5$$
$$10 = 3 + 7$$
$$12 = 5 + 7$$
$$14 = 7 + 7$$

3   Some integers can be written as a sum of two primes in more than one way. For example, $10 = 3 + 7$ and $10 = 5 + 5$.

4   There are many variants of the Goldbach conjecture. We list them below:

- **Strong Goldbach 1**: Each even integer greater than 2 is the sum of two primes.

4.5 Goldbach conjecture

- **Strong Goldbach 2**: Each even integer greater than 4 is the sum of two odd primes.
- **Strong Goldbach 3**: Each integer greater than 5 is the sum of three primes.
- **Weak Goldbach 1**: Each odd integer greater than 5 is the sum of three primes.
- **Weak Goldbach 2**: Each odd integer greater than 7 is the sum of three odd primes.

5 In this section, we show that all the strong Goldbach conjectures are equivalent. Then we show that each weak Goldbach conjecture is implied by each strong Goldbach conjecture.

Before this, we prove some lemmas that help us understand the relationship between the variants of the conjecture

6 **Lemmas for Goldbach conjecture.** Here we prove the following lemmas.

- Sum of three even integers is even
- Sum of three odd integers is odd
- If the sum of three integers is odd then all the integers are odd or two integers are even and one is odd.

7 The sum of three even integers is even.

Proof Let a, b, c be even integers. Thus $a + b$ is even. Since $a + b$ and c are even, it follows that their sum $(a + b) + c$ is even. Hence $a + b + c$ is even. ∎

8 The sum of three odd integers is odd

Proof Let a, b, c be three odd integers. Thus,

$$a = 2k_1 - 1$$
$$b = 2k_2 - 1$$
$$c = 2k_3 - 1$$

for some integers k_1, k_2, k_3. Hence,

$$a + b + c = (2k_1 - 1) + (2k_2 - 1) + (2k_3 - 1)$$
$$= (2k_1 + 2k_2 + 2k_3) - 1 - 1 - 1$$
$$= 2(k_1 + k_2 + k_3) - 2 - 1$$

$$= 2(k_1 + k_2 + k_3 - 1) - 1$$
$$= 2k_4 - 1 \quad \text{for some integer } k_4$$

Therefore $a + b + c$ is odd. ∎

9 If the sum of three integers is odd then all the integers are odd or two integers are even and one is odd.

Proof The number n of even integers in the list is 0,1,2 or 3.

When $n = 0$

In this case, all three integers are odd. Hence $a + b + c$ is odd.

When $n = 1$

In this case, there is exactly one even integer. The sum of both odd integers is even. If we add the result of this sum to the only even integer in the list, we get an even integer. Thus $a + b + c$ is even.

When $n = 2$

Here, there are exactly two even integers. Let a, b be the even integers. Let c be the odd integer. Hence $a + b$ is even. Since $a + b$ is even and c is odd, it follows that $a + b + c$ is odd.

When $n = 3$

Here, all three integers are even. Thus their sum is even.

Conclusion

In all cases, the sum is odd if and only if $n = 0$ or $n = 2$. Therefore the statement is true. ∎

10 **Relationship between the variants of the Goldbach conjecture.** In this section we show the relationship between variants of the Goldbach conjecture.

11 Strong Goldbach 1 ↔ 2

Each even integer greater than 2 is the sum of two primes if and only if each even integer greater than 4 is the sum of two odd primes.

Proof

Part 1: If each even integer greater than 2 is the sum of two primes then each even integer greater than 4 is the sum of two odd primes.
Let n be an even integer greater than 4. Thus

$$n > 4 \qquad (1)$$

Hence $n > 2$. By Strong Goldbach 1, $n = p_1 + p_2$ for some primes p_1, p_2.

It is either p_1, p_2 are both odd or they are both even. We prove that they are not both even using contradiction. Assume they are both even. Then $p_1 = p_2 = 2$. Hence $n = 4$. This contradicts with (1). Hence p_1, p_2 are not both even. By a previous lemma, they are both odd. Hence n is a sum of two odd primes.

Part 2: If each even integer greater than 4 is the sum of two odd primes then each even integer greater than 2 is the sum of two primes.
Let n be an even integer greater than 2. Thus $n = 4$ or $n > 4$. In the first case $n = 4$.

Observe that
$$4 = 2 + 2$$

Thus n is a sum of two primes. In the second case $n > 4$. By Strong Goldbach 2, n is the sum of two odd primes. Thus n is the sum of two primes. ∎

12 **Strong Goldbach 1 ↔ 3**

Each even integer greater than 2 is the sum of two primes if and only if each integer greater than 5 is the sum of three primes.

Proof

Part 1: If even integer greater than 2 is the sum of two primes then each integer greater than 5 is the sum of three primes.
Let n be an integer greater than 5. It is either n is even or n is odd. If n is even then $n \geq 6$. Thus $n - 2$ is an even integer greater than 2. By Strong Goldbach 1, $n - 2$ is the sum of two primes. Hence, for some primes p_1, p_2,

$$n - 2 = p_1 + p_2$$
$$n = 2 + p_1 + p_2$$

Since $2, p_1, p_2$ are all primes, it follows that n is the sum of three primes.

If n is odd then $n \geq 7$. Thus $n - 3 \geq 4$. Hence $n - 3$ is an even integer greater than 2. By Strong Goldbach 1, $n - 3$ is a sum of two primes. Thus, for some primes p_1, p_2,

$$n - 3 = p_1 + p_2$$
$$n = 3 + p_1 + p_2$$

Since $3, p_1, p_2$ are all primes, it follows that n is the sum of three primes.

Part 1: If each integer greater than 5 is the sum of three primes then each even integer greater than 2 is the sum of two primes Let n be an even integer greater than 2. Thus $n > 2$. It follows that $n + 2 > 4$. Since $n + 2$ is an even integer, $n + 2 \geq 6$. Hence $n + 2 > 5$. By Strong Goldbach 3, $n + 2$ is the sum of three primes. Hence, for some primes a, b, c,

$$n + 2 = a + b + c$$

Since $a + b + c$ is even, by a previous theorem, it is either a, b, c are all even or two of the primes are even and one is odd. If they are all even then $a = b = c = 2$. Thus

$$n + 2 = 2 + 2 + 2$$
$$n = 2 + 2$$

Thus n is the sum of two primes.

If two of the primes are odd and one is even then one of the primes is 2. Let $a = 2$. Then

$$n + 2 = 2 + b + c$$
$$n = b + c$$

Hence n is the sum of two primes.

∎

13 Strong Goldbach 1 → Weak Goldbach 2

If each even integer greater than 2 is the sum of two primes then each odd integer greater than 7 is the sum of three odd primes

Proof Let n be an odd integer greater than 7. Thus $n > 7$. Hence $n - 5 > 2$.

Thus $n - 5$ is an even integer greater than 2. By Strong Goldbach 1, $n - 5$ is the sum of two primes. It is either both primes are even or both primes are odd. If both primes are even then the primes are both equal to 2. Thus

$$n - 5 = 2 + 2$$
$$n = 2 + 2 + 5$$
$$= 3 + 3 + 3$$

Thus n is the sum of three odd primes.

If both primes are odd then for some odd primes a, b

$$n - 5 = a + b$$
$$n = 5 + a + b$$

Hence n is the sum of three odd primes. ∎

14 Weak Goldbach 2 → Weak Goldbach 1

If each odd integer greater than 7 is the sum of three odd primes then each odd integer greater than 5 is the sum of three primes

Proof Let n be an odd integer greater than 5. Thus $n \geq 7$

Hence $n = 7$ or $n > 7$. If $n = 7$ then $n = 2 + 2 + 3$. Thus n is the sum of three primes. If $n > 7$ then, by weak Goldbach 2, n is the sum of three odd primes. Thus n is the sum of three primes. In both cases, n is the sum of three primes. ∎

Exercises 4.5

Prove each of the following.

1 Strong Goldbach 1 implies Strong Goldbach 2

2 Strong Goldbach 1 implies Weak Goldbach 1

3 Strong Goldbach 2 implies Weak Goldbach 1

4 Strong Goldbach 2 implies Weak Goldbach 2

4.6 Forms of primes

1 In this section, we use the division algorithm to classify the primes into forms. Then we show that

- there are infinitely many primes of the form $4n - 1$
- there are infinitely many primes of the form $6n - 1$

2 **Forms of primes according to the division algorithm.** Here, we use the division algorithm to show that

- each prime greater than 3 is of the form $3n + 1$ or $3n - 1$
- each prime greater than 2 is of the form $4n + 1$ or $4n - 1$
- each prime greater than 3 is of the form $6n + 1$ or $6n - 1$
- each prime greater than 5 is of the form $30n \pm 1$, $30n \pm 7$, $30n \pm 11$ or $30n \pm 13$.

Before this, we prove some lemmas:

- For all positive integers n, m, if $n + m$ is prime then $\gcd(n, m) = 1$.
- Let b be an integer greater than 1. For each prime number p, if $p \nmid b$ then for some integers q, r,

$$p = bq + r, \quad 0 < r < b \quad \text{and} \quad \gcd(b, r) = 1.$$

3 For all positive integers n, m, if $n + m$ is prime then $\gcd(n, m) = 1$.

Proof We prove the statement by contradiction. Assume $\gcd(n, m) \neq 1$. Thus $\gcd(n, m) > 1$. Let $d = \gcd(n, m)$. Thus $d > 1$ and $d \mid n$, $d \mid m$. Hence $d \mid n + m$. Since $n + m$ is prime, its positive divisors are 1 and itself. Since $d \neq 1$, it follows that $d = n + m$. Hence,

$$n + m \mid n$$
$$n + m \leq n$$

Subtract n from both sides

$$m \leq 0$$

This is absurd because $m > 0$. Thus $\gcd(n, m) = 1$. ∎

4 Let b be an integer greater than 1. For each prime number p, if $p \nmid b$ then for some integers q, r,

$$p = bq + r, \quad 0 < r < b \quad \text{and} \quad \gcd(b, r) = 1.$$

Proof Let b be an integer greater than 1. Let p be a prime number such that $p \nmid b$. Thus for all integers q,

$$p \neq bq$$
$$p - bq \neq 0 \tag{1}$$

By the division algorithm, there exist integers q, r such that

$$p = bq + r \quad \text{and} \quad 0 \leq r < b.$$

Since $p - bq \neq 0$, it follows that $r \neq 0$. Thus

$$p = bq + r \quad \text{and} \quad 0 < r < b$$

Since $p > 0$, it follows that $bq + r > 0$. Hence $bq > -r$. Since $r < b$, it follows that $-r > -b$. Thus $bq > -r$ and $-r > -b$. Hence $bq > -b$.ND both sides by b we get $q > -1$. Hence $q \geq 0$. It is either $q = 0$ or $q > 0$.

When $q = 0$: In this case, $q = 0$. Thus $p = b(0) + r$. Hence $p = r$. Since $p \nmid b$, $\gcd(b, p) = 1$. Thus $\gcd(b, r) = 1$. Hence $p = bq + r$, $0 < r < b$ and $\gcd(b, r) = 1$.

When $q > 0$: In this case $q > 0$. Thus $bq > 0$ and $r > 0$. Hence $bq + r$ is the sum of two positive integers. Since $bq + r$ is prime, $\gcd(bq, r) = 1$. Thus $p = bq + r$, $0 < r < b$ and $\gcd(b, r) = 1$.

Conclusion In both cases, we have that $p = bq + r$, $0 < r < b$ and $\gcd(b, r) = 1$. ∎

5 Each prime greater than 2 is of the form $2n - 1$.

Proof Let p be a prime greater than 2. Thus $p \neq 2$. Since 2 is the only even prime, it follows that p is odd. Thus p is of the form $2n - 1$. ∎

6 Each prime greater than 3 is of the form $3n + 1$ or $3n - 1$.

Proof Let p be a prime greater than 3. Thus $p \nmid 3$. By a previous theorem, $p = 3q + r$ and $0 < r < 3$ and $\gcd(3, r) = 1$ for some integers q, r. Since $0 < r < 3$, it follows that $r = 1$ or $r = 2$. Thus $p = 3q + 1$ or $p = 3q + 2$. Hence p is of the form $3n + 1$ or $3n + 2$. Each integer of the form $3n + 2$ is of the form $3n - 1$. Thus p is of the form $3n + 1$ or $3n - 1$. ∎

7 Each prime greater than 2 is of the form $4n + 1$ or $4n - 1$

Proof Let p be a prime greater than 2. Thus $p \nmid 2$. Hence $p \nmid 4$. By a previous theorem, $p = 4q + r$, $0 < r < 4$ and $\gcd(4, r) = 1$ for some integers q, r. Since $0 < r < 4$, it follows that $r = 1, 2, 3$. Since $\gcd(4, r) = 1$, it follows that $r \neq 2$. Hence $r = 1$ or $r = 3$. Thus $p = 4q + 1$ or $p = 4q + 3$. Hence p is of the form $4n + 1$ or of the form $4n + 3$. Each integer of the form $4n + 3$ is also of the form $4n - 1$. Thus p is of the form $4n + 1$ or of the form $4n - 1$. ∎

8 Each prime greater than 3 is of the form $6n + 1$ or $6n - 1$

Proof Let p be a prime greater than 3. Thus $2, 3 \neq p$. Hence $p \nmid 2 \cdot 3$. Thus $p \nmid 6$. By a previous theorem, $p = 6q + r$, $0 < r < 6$ and $\gcd(6, r) = 1$ for some integers q, r. Since $0 < r < 6$, it follows that r is in the list $1, 2, 3, 4, 5$. Since $\gcd(6, r) = 1$, it follows that $2, 3, 4 \neq r$. Thus r is in the list $1, 5$. Hence $r = 1$ or $r = 5$. Thus $p = 6q + 1$ or $p = 6q + 5$. Therefore p is of the form $6n + 1$ or $6n + 5$. Each prime of the form $6n + 5$ is of the form $6n - 1$. Hence p is of the form $6n + 1$ or $6n - 1$. ∎

9 Each prime greater than 5 is of the form $30n \pm 1$, $30n \pm 7$, $30n \pm 11$ or $30n \pm 13$.

Proof Let p be a prime number greater than 5. The prime factors of 30 are $2, 3, 5$. Since $p > 5$, it follows that $p \nmid 2 \cdot 3 \cdot 5$. Hence $p \nmid 30$. By a previous theorem, $p = 30q + r$, $0 < r < 30$ and $\gcd(30, r) = 1$ for some integers q, r. Since $0 < r < 30$, it follows that r is in the list $1, \ldots, 29$. Since $\gcd(30, r) = 1$, it follows that r is a positive integer relatively prime to 30 and less than 30. The integers in $1, \ldots, 29$ that are relatively prime to 30 are

$$1, 7, 11, 13,$$

$$17, 19, 23, 29$$

Thus $p = 30q + r$ where $r = 1, 7, 11, 13, 17, 19, 23$ or 29. Hence p is of the form $30n + 1$, $30n + 7$, $30n + 11$, $30n + 13$, $30n + 17$, $30n + 19$, $30n + 23$ or $30n + 29$.

Observe that the forms $30n+17, 30n+19, 30n+23$ and $30n+29$ are equivalent to the forms $30n - 13, 30n - 11, 30n - 7$ and $30n - 1$ respectively. Therefore p is of the form

$$30n + 1, 30n + 7, 30n + 11, 30n + 13,$$

$$30n - 1, 30n - 7, 30n - 11, 30n - 13,$$

Therefore p is of one of the forms below

$$30n \pm 1, 30n \pm 7, 30n \pm 11, 30n \pm 13$$

■

10 We have shown that each prime greater than 3 is of the form $6n + 1$ or $6n - 1$. Below we prove a corollary of this theorem.

11 Each prime of the form $3n + 1$ is also of the form $6n + 1$.

Proof Let p be a prime of the form $3n + 1$. Since 2 and 3 are each not of the form $3n + 1$, it follows that $p > 3$. Hence, by a previous theorem, p is of the form $6n + 1$ or $6n - 1$.

Since p is a prime of the form $3n + 1$, by a previous theorem, p is not of the form $3n - 1$. Each integer of the form $6n - 1$ is of the form $3n - 1$. Thus p is not of the form $6n - 1$. Therefore p is of the form $6n + 1$.

■

12 **Infinitude of primes of a certain form.** Here, we show that

- there are infinitely many primes of the form $4n - 1$
- there are infinitely many primes of the form $6n - 1$

In order to prove these theorems, we need two lemmas:

- Each integer of the form $ak - 1$ is not of the form $ak + 1$.
- If b is not of the form $an + 1$ then at least one of its prime factors is not of the form $an + 1$.

13 Let a be an integer greater than 2. Each integer of the form $ak - 1$ is not of the form $ak + 1$.

Proof Let n be an integer of the form $ak - 1$. Thus

$$n = aj - 1$$

for some integer j

Assume n is of the form $ak + 1$. Thus, for some integer k,

$$n = ak + 1$$

Hence

$$ak + 1 = aj - 1$$
$$1 + 1 = aj - ak$$
$$2 = a(j - k)$$

Hence

$$a \mid 2$$
$$a \leq 2$$

This is absurd because $a > 2$. Hence n is not of the form $ak + 1$. ∎

14 Let a, n be positive integers. The product of n integers of the form $ak + 1$ is of the form $ak + 1$.

Proof We prove the statement by induction. For the base case, let $n = 1$. The product of 1 integer of the form $ak + 1$ is of the form $ak + 1$. Thus we have established the base case. For the induction step, assume the statement is true at $n = m$, where m is a positive integer. Thus the product of m integers of the form $ak + 1$ is of the form $ak + 1$. Let $b_1, \ldots, b_m, b_{m+1}$ be $m + 1$ integers of the form $ak + 1$. Let $N = b_1 \cdots b_m$. By the induction hypothesis, N is of the form $ak + 1$. Thus, for some integer j_1

$$N = a \cdot j_1 + 1 \tag{1}$$

Since b_{m+1} is of the form $ak + 1$, for some integer j_2,

$$b_{m+1} = aj_2 + 1 \tag{2}$$

Multiply (1) and (2)

$$N \cdot b_{m+1} = (aj_1 + 1)(aj_2 + 1)$$

4.6 Forms of primes

$$= aj_1 \cdot aj_2 + aj_1 + aj_2 + 1$$
$$= a(j_1 \cdot j_2 + j_1 + j_2) + 1$$
$$= aj_3 + 1$$

for some integer j_3.

Hence $N \cdot b_{m+1}$ is of the form $ak + 1$. Thus $b_1 \cdots b_m \cdot b_{m+1}$ is of the form $ak + 1$. Therefore the product of $m + 1$ integers of the form $ak + 1$ is of the form $ak + 1$. This establishes the induction step. Therefore the statement is true. ∎

15 Let a, b be integers greater than 1. If all prime factors of b are of the form $an + 1$ then b is also of the form $an + 1$.

Proof Since $b > 1$, it follows that b has a prime factorization

$$b = p_1 \cdots p_m$$

for some primes $p_1, \ldots, p_m$. Since $p_1, \ldots, p_m$ are prime factors of b, they are of the form $an + 1$. Thus their product $p_1 \cdots p_m$ is also of the form $an + 1$. Hence b is of the form $an + 1$. ∎

16 Let a, b be integers greater than 1. If b is not of the form $an + 1$ then at least one of its prime factors is not of the form $an + 1$.

Proof This is the contrapositive of the previous theorem. ∎

17 Let n be an integer greater than 1. If n is of the form $4k - 1$ then n has a prime factor of the form $4k - 1$.

Proof Let n be an integer greater than 1. Let n be of the form $4k - 1$. Hence $n = 4j - 1$ for some integer j. Since $2 \mid 4j$, it follows that $2 \nmid 4j - 1$. Hence $2 \nmid n$. Thus 2 is not a prime factor of n. Hence all prime factors of n are greater than 2. Since n is of the form $4k - 1$, n is not of the form $4k + 1$. Thus n has a prime factor p which is not of the form $4k + 1$. Since $p \mid n$, it follows that $p > 2$. Thus p is of the form $4k + 1$ or $4k - 1$. We know that p is not of the form $4k + 1$. Thus p is of the form $4k - 1$. Hence n has a prime factor of the form $4k - 1$. ∎

18 Let n be an integer greater than 1. If n is of the form $6k - 1$ then n has a prime factor of the form $6k - 1$.

Proof Let n be an integer greater than 1. Let n be of the form $6k - 1$. Hence $n = 6j - 1$ for some integer j. Since $2 \mid 6j$, it follows that $2 \nmid 6j - 1$. Hence $2 \nmid n$. Since $3 \mid 6j$, it follows that $3 \nmid 6j - 1$. Hence $3 \nmid n$. Thus 2 and 3 are not prime factors of n. Since n is of the form $6k - 1$, n is not of the form $6k + 1$. Thus n has a prime factor p which is not of the form $6k + 1$. Since 2 and 3 are not prime factors of n, it follows that $2, 3 \neq p$. Hence $p > 3$. Thus p is either of the form $6k + 1$ or of the form $6k - 1$. Since p is not of the form $6k + 1$, it follows that p is of the form $6k - 1$. Hence n has a prime factor of the form $6k - 1$. ∎

19 There are infinitely many primes of the form $4n - 1$

Proof We prove that given a finite list of primes of the form $4n - 1$, there exists a prime of the form $4n - 1$ that is not in the list. Let $p_1, \ldots, p_m$ be a list of primes of the form $4n - 1$. Let $q = p_1 \cdots p_m$. Hence $4q - 1 > 1$. Thus $4q - 1$ has a prime factor p of the form $4n - 1$. Since $p \mid 4q - 1$, it follows that $p \nmid 4q$. Hence $p \nmid q$. Thus $p \nmid p_1 \cdots p_m$. Hence p is not in the list $p_1, \ldots, p_m$. Therefore there exists a prime of the form $4n - 1$ that is not in the list $p_1, \ldots, p_m$. Therefore there are infinitely many primes of the form $4n - 1$. ∎

20 There are infinitely many primes of the form $6n - 1$

Proof Let $p_1, \ldots, p_m$ be a list of primes of the form $6n - 1$. Let $q = p_1 \cdots p_m$. Hence $6q - 1 > 1$.

Since $6q - 1$ is an integer greater than 1, it follows that $6q - 1$ has a prime factor p of the form $6n - 1$. Since $p \mid 6q - 1$, it follows that $p \nmid 6q$. Hence $p \nmid q$. Thus $p \nmid p_1 \cdots p_m$. Hence p is not in the list $p_1, \ldots, p_m$. Thus there exists a prime of the form $6n - 1$ that is not in the list $p_1, \ldots, p_m$. Therefore every finite list of primes of the form $6n - 1$ is incomplete. Therefore there are infinitely many primes of the form $6n - 1$. ∎

CHAPTER 5

Primes in AP

5.1 Introduction

1 In this section, we define primes in AP. Then we discuss the relationship between integers in AP and forms of integers. We prove the following theorems:

- no nonconstant AP consists solely of primes
- we can find consecutive composite terms of any length in an AP provided that the first term and common difference are coprime
- if n consecutive terms of an AP are prime numbers then the common difference is divisible by every prime less than n.

Then we discuss Green-Tao theorem.

2 **Definition of primes in AP.** A sequence of primes in AP is a list of three or more primes which are consecutive terms of some AP.

3 Let us consider an example of a sequence of primes in AP. Consider the AP $a_0, a_1, \ldots$ defined thus: for all nonnegative integers n,

$$a_n = 7 + 30n$$

Hence

$$a_0 = 7$$
$$a_1 = 7 + 30 = 37$$
$$a_2 = 7 + 30(2) = 67$$
$$a_3 = 7 + 30(3) = 97$$
$$a_4 = 7 + 30(4) = 127$$
$$a_5 = 7 + 30(5) = 157$$

The integers $a_0, \ldots, a_5$ are primes. Since they are consecutive terms of an AP, we say the list $a_0, \ldots, a_5$ is a sequence of primes in AP. Observe that

$$a_6 = 7 + 30(6)$$
$$= 11 \cdot 17$$

Hence a_6 is composite. Thus not all integers in the AP are prime.

4   Other examples of primes in AP are given below

1. $3, 5, 7$ which is $3 + 2n$ for $0 \leq n \leq 2$
2. $5, 11, 17, 23, 29$ which is $5 + 6n$ for $0 \leq n \leq 4$

5   **Integer forms and integers in AP.** There is a relationship between the form of an integer and whether it is in an AP. For example, if an integer is in the AP $1, 1 + 2, 1 + 2(2), \ldots$ then it is of the form $1 + 2n$. On the other hand, if an integer is of the form $1 + 2n$ then it is either in the AP $1, 1 + 2, 1 + 2(2), \ldots$ or in the AP $1 - 2, 1 - 2(2), \ldots$.

6   We prove this theorem below.

7   Let a, b be positive integers. Then an integer is of the form $a + bn$ if and only if the integer is in the AP $a, a + b, a + 2b, \ldots$ or in the AP $a - b, a - 2b, \ldots$

Proof Let n be an integer.

Part 1: If n of the form $a + bn$ then n is in the AP $a, a + b, a + 2b, \ldots$ or in the AP $a - b, a - 2b, \ldots$ Let n be of the form $a + bj$. Thus $n = a + bj$ for some integer j. Since j is an integer, it follows that $j < 0$ or $j \geq 0$. If $j < 0$ then j is in the sequence $-1, -2, \ldots$. Thus n is in the AP $a - b, a - 2b, \ldots$. If $j \geq 0$ then j is in the sequence $0, 1, 2, \ldots$. Thus n is in the AP $a, a + b, a + 2b, \ldots$. Hence in both cases n is in the AP $a - b, a - 2b, \ldots$ or in the AP $a, a + b, a + 2b, \ldots$.

Part 2: If n is in the AP $a, a + b, a + 2b, \ldots$ or in the AP $a - b, a - 2b, \ldots$ then n is of the form $a + bn$.   Let n be an integer in the AP $a - b, a - 2b, \ldots$. Then n is of the form $a - bj$. Hence n is of the form $a + bj$

Let n be an integer in the AP $a, a + b, a + 2b, \ldots$. Then n is of the form $a + bj$.

Thus in both cases n is of the form $a + bj$.

∎

8 **AP with infinitely many composite integers.** Here we show that there are infinitely many composite integers in each nonconstant AP of integers greater than 1. Thus no nonconstant AP consists solely of primes.

9 Each nonconstant AP of integers greater than 1 contains infinitely many composite integers.

Proof Let f be a nonconstant AP of integers greater than 1. Thus for all positive integers n,
$$f(n) = a + bn$$
for some integers a, b.

Since the AP is nonconstant, the common difference is not zero. Hence,
$$b \neq 0 \tag{1}$$

For all integers $n > 0$, let
$$g(n) = f(1)n + 1$$

Hence, for all integers $n > 0$,
$$\begin{aligned} f(g(n)) &= a + bg(n) \\ &= a + b(f(1)n + 1) \\ &= a + bf(1)n + b \\ &= (a + b) + bnf(1) \\ &= f(1) + bnf(1) \\ &= f(1)(1 + bn) \end{aligned}$$

Since $f(g(n))$ and $f(1)$ are members of the AP, they are positive integers. Hence
$$\frac{f(g(n))}{f(1)} > 0$$
$$1 + bn > 0$$
$$bn + 1 \geq 1$$
$$bn \geq 0$$

Since b and n are nonzero, their product bn is nonzero. Thus
$$bn > 0$$
$$bn + 1 > 1$$

Hence $f(1)$ and $bn+1$ are integers greater than 1. Thus their product $f(1)(bn+1)$ is a composite integer. Therefore $f(g(n))$ is a composite integer. Thus the integers $f(g(1)), f(g(2)), \ldots$ are composite integers. Each of these integers are in the AP. Thus the AP contains infinitely many composite integers. ∎

10 No nonconstant AP of integers greater than 1 consists solely of prime numbers.

Proof Let $a_1, a_2, \ldots$ be a nonconstant AP of integers greater than 1. By a previous theorem, there exist infinitely many composite integers in the AP. Hence there exists a composite integer in the AP. Therefore the AP does not consist solely of primes. ∎

11 **Consecutive composite terms in an AP.** Here we show that we can find consecutive composite terms of any length in an AP provided that the first term and common difference are coprime.

12 Let a, b, k be positive integers. The arithmetic progression

$$a + b, a + 2b, a + 3b, \ldots$$

where $\gcd(a, b) = 1$, contains k consecutive terms that are composite.

Proof Let $f(n)$ be an arithmetic progression defined thus

$$f(n) = a + nb$$

Let $g(n)$ be a sequence of positive integers defined thus

$$g(n) = f(1)f(2) \cdots f(n)$$

for all positive integers n. Now consider the integer

$$f(g(n) + k)$$

for some integer k.

$$\begin{aligned} f(g(n) + k) &= a + [g(n) + k]b \\ &= a + g(n)b + kb \\ &= (a + kb) + bg(n) \end{aligned} \quad (1)$$

Now, if $1 \leq k \leq n$ then $a + kb$ is among the list $(a+b), (a+2b), \ldots (a+nb)$. Thus

$$(a + kb) \mid (a + b)(a + 2b) \cdots (a + nb)$$

$$(a + kb) \mid g(n)$$
$$(a + kb) \mid bg(n)$$
$$bg(n) = (a + kb)m \quad \text{for some integer } m \tag{2}$$

since $bg(n)$ and $a + kb$ are both positive then

$$m > 0$$

add 1 to both sides

$$1 + m > 1 \tag{3}$$

from (2) we can write (1) as

$$f(g(n) + k) = (a + kb) + (a + kb)m$$
$$f(g(n) + k) = (a + kb)(1 + m) \tag{4}$$

We know that

$$b \geq 1 \tag{5}$$
$$k \geq 1 \tag{6}$$
$$a \geq 1 \tag{7}$$

multiply (5) and (6) together

$$bk \geq 1 \tag{8}$$

add (7) and (8) together

$$a + bk \geq 2$$
$$a + bk > 1 \tag{9}$$

From (3), (9) and (4) we can conclude that $f(g(n) + k)$ is composite for all $1 \leq k \leq n$. Hence

$$f(g(n) + 1), f(g(n) + 2), \ldots, f(g(n) + n)$$

are n consecutive terms of the AP that are composite.

∎

13 **Common difference of primes in AP.** Here we show that if n consecutive terms of an arithmetic progression are prime numbers then the common difference d is divisible by every prime less than n.

Before we prove this theorem, we prove a lemma below.

14 Let a, b, n be positive integers. If $\gcd(b, n) = 1$ then one of the integers

$$a, a + b, \ldots, a + (n - 1)b$$

is divisible by n.

Proof Since $\gcd(b, n) = 1$, it follows that 1 is a linear combination of b and n. Hence, $1 = nk + bs$ for some integers k, s. Thus

$$nk = 1 - bs \qquad (1)$$

Since $-as$ and n are integers, and $n \neq 0$, by the division algorithm, $-as = nq + r$ and $0 \leq r < n$ for some integers q, r.

Thus,

$$r = -as - nq$$

Multiply both sides by b

$$br = b(-as - nq)$$

Add a to both sides

$$a + br = a + b(-as - nq)$$
$$= a - bas - bnq$$
$$= a(1 - bs) - bnq$$

From (1), we get

$$= ank - bnq$$
$$= n(ak - bq)$$

Thus $n \mid a + br$.

Since $0 \leq r < n$, it follows that r is in $0, \ldots, n - 1$. Thus $a + br$ is in

$$a + b(0), \ldots, a + b(n - 1)$$

Hence one of the integers in $a + b(0), \ldots, a + b(n - 1)$ is divisible by n. ∎

15 Let $n > 1$ and let $d > 0$. If $p, p + d, \ldots, p + (n - 1)d$ are prime numbers then d is divisible by every prime less than n.

Proof We prove this statement by contradiction. Let q be a prime less than n. Assume $q \nmid d$. Thus $\gcd(q, d) = 1$. By a previous theorem, one of the integers $p, p+d, \ldots, p+(q-1)d$ is divisible by q. Hence q is in the list $p, \ldots, p+(q-1)d$. Thus for some $0 \leq i \leq q - 1$,

$$q = p + id$$

Since $i \geq 0$ and $d \geq 0$, we get

$$id \geq 0$$
$$q - p \geq 0$$
$$q \geq p$$

Since $q < n$, it follows that $n > q$ and $q \geq p$. Thus $n > p$. Hence $n - 1 \geq p$. Thus $0 \leq p \leq n - 1$. Hence p is in $0, \ldots, n - 1$. Thus $p + pd$ is in $p + d(0), \ldots, p + p(n-1)$. Observe that $p > 1$ and $d + 1 > 1$. Thus $p(d+1)$ is a composite integer. Hence the list $p, \ldots, p + p(n-1)$ contains a composite integer. This is absurd because the list consists solely of primes. Hence, $q \mid d$. It follows that every prime less than n divides d. ∎

16 **Green-Tao theorem.** The examples of primes in AP we have considered have just a few primes in AP. Can we find a sequence of more than 100 primes in AP? What is the maximum length for a sequence of primes in AP? The Green-Tao theorem answers these questions.

17 According to Green-Tao theorem there is no maximum length. Thus there exist arbitrarily long sequences of primes in AP. We state the Green-Tao theorem below.

18 For each positive integer n, there exists a sequence of more than n primes in AP.

(Green-Tao Theorem)

19 The proof of this theorem is beyond the scope of this book.

Exercises 5.1

For each integer n greater than 1, the primorial of n is the product of all the primes not greater than n. We denote the primorial of n by $n\#$. For example,

$$2\# = 2$$

$$3\# = 2 \cdot 3$$
$$4\# = 2 \cdot 3$$
$$5\# = 2 \cdot 3 \cdot 5$$
$$6\# = 2 \cdot 3 \cdot 5$$

Let $n > 1$ and let $d > 0$. Prove that if $p, p+d, \ldots, p+(n-1)d$ are prime numbers then d is divisible by $n\#$.

5.2 Dirichlet's theorem

1 In this section, we state Dirichlet's theorem and use it to prove the following: there are infinitely many primes

   1. ending in 33
   2. ending in as many 1's as required

2 Dirichlet's theorem states that if a and b are relatively prime positive integers, then the arithmetic progression

$$a, a+b, a+2b, \ldots$$

contains infinitely many primes.

The proof of this theorem is too complex at this level.

3 **Primes ending in** 33. Examples of primes ending in 33 are

$$233,$$
$$433,$$
$$733,$$
$$1033$$

Just to be clear, an integer that ends in 333 also ends in 33. Hence in this section we are concerned about primes that end in *at least* two threes in their decimal representation.

4 There exist infinitely many primes ending in 33.

Proof Consider the AP whose nth term $f(n)$ is defined thus

$$f(n) = 100n + 33$$

5.2 Dirichlet's theorem

We have that the terms in this AP always end in 33. Since $\gcd(100, 33) = 1$ then Dirichlet's theorem gives that there are infinitely many primes in this A.P. Hence there are infinitely many primes that end in 33. ∎

5 **Primes ending in 1's.** Here we prove that there exists a prime ending in as many consecutive 1's as required. Before this, we prove the following lemmas.

- Let n be a positive integer. Each prime factor of 10^n is 2 or 5.
- For all positive integers m, n,
$$\gcd(10^n, 10^m - 1) = 1$$

6 Let n be a positive integer. Each prime factor of 10^n is 2 or 5.

Proof Let p be a prime factor of 10^n. Thus $p \mid 10^n$. Since $10^n = 2^n \cdot 5^n$, it follows that $p \mid 2^n \cdot 5^n$. Hence $p \mid 2^n$ or $p \mid 5^n$. Thus $p = 2$ or $p = 5$. ∎

7 For all positive integers m, n,
$$\gcd(10^n, 10^m - 1) = 1$$

Proof Since $2 \nmid 10^m - 1$ and $5 \nmid 10^m - 1$ it follows that 2 and 5 are not prime factors of $10^m - 1$. The only prime factors of 10^n are 2 and 5. Hence 10^n and $10^m - 1$ have no prime factors in common. Therefore $\gcd(10^n, 10^m - 1) = 1$. ∎

8 There exists a prime ending in as many consecutive 1's as required.

Proof Let
$$R_n = \frac{10^n - 1}{9}$$
for all positive integers n. Thus
$$R_1 = 1, R_2 = 11, \ldots, R_n = \underbrace{111 \cdots 1}_{n}$$

Let k be a positive integer. Consider the AP $f(n)$ defined thus
$$f(n) = 10^k n + R_n$$
for all positive integers n. All terms in the AP end in at least k 1's. Now $\gcd(10^k, 10^n - 1) = 1$. Hence $\gcd(10^k, 9 \cdot R_n) = 1$. Thus $\gcd(10^k, R_n) = 1$.

Since $\gcd(10^k, R_n) = 1$, by Dirichlet's theorem, the AP contains infinitely many primes. Hence there are infinitely many primes ending in at least k 1's for $k = 1, 2, \ldots$.

∎

Exercises 5.2

Prove the following.

1 Each prime is contained in infinitely many APs.

2 Let a, b, n be positive integers such that $1 < a < n$. Then one of the following integers is composite: $a, a+b, \ldots, a+b(n-1)$

CHAPTER 6

Prime gaps

6.1 Twin primes

1. In this section, we define twin prime and state the twin prime conjecture. We prove some properties of twin primes. Then we discuss isolated primes and prime triplets.

2. Before discussing twin primes let us discuss the smallest gap possible between primes.

3. **Smallest prime gap.** The smallest gap between primes is one. This occurs only once with the primes 2 and 3. We prove this below.

4. Let a be a positive integer. If $p, p + 1$ are primes then $p = 2$.

   Proof One of the integers $p, p + 1$ is even. Since 2 is the only even prime, it follows that one of the primes is 2. Hence $p = 2$ or $p + 1 = 2$. Thus $p = 2$ or $p = 1$. Since 1 is not prime, it follows that $p = 2$. ∎

5. **Definition of twin prime.** In the sequence of primes, we come across prime pairs that differ by 2. Examples are 3 and 5, 101 and 103. Such pairs of primes are called twin primes. A twin prime is a prime p such that $p + 2$ or $p - 2$ is prime.

6. The first few pairs of twin primes are (3, 5), (5, 7), (11, 13), (17, 19), (29, 31), (41, 43).

7. The twin prime conjecture states that there are infinitely many twin primes. Researchers have found over 800 trillion twin prime pairs below 10^{18}.

8. **Properties of twin primes.** Here, we prove the following:

   1. the smaller prime in a pair of twin primes is of the form $6n - 1$

2. the sum of twin primes is divisible by 12

9. Let p, $p + 2$ be primes greater than 3. Then p is of the form $6n - 1$.

   Proof Since p is a prime greater than 3 it follows that p is of the form $6n + 1$ or $6n - 1$. We prove by contradiction that p is of the form $6n - 1$. Assume p is not of the form $6n - 1$. Hence

   $$p = 6j + 1 \quad \text{for some integer } j$$
   $$p + 2 = 6j + 1 + 2$$
   $$= 6j + 3$$
   $$= 3(2j + 1)$$

   Thus $3 \mid p + 2$. Since $p + 2$ is prime it follows that $p + 2 = 3$. This is absurd because $p + 2$ is a prime greater than 3. Thus p is of the form $6n - 1$. ∎

10. Let p, $p + 2$ be primes greater than 3. Then their sum $p + (p + 2)$ is divisible by 12.

    Proof By a previous theorem, we know that p is of the form $6n - 1$. Hence,

    $$p = 6j - 1 \quad \text{for some integer } j$$
    $$p + 2 = 6j + 1$$
    $$p + (p + 2) = (6j - 1) + (6j + 1)$$
    $$= 6j + 6j$$
    $$= 12j$$

    Thus $12 \mid p + (p + 2)$ ∎

11. **Isolated primes.** The opposite of a twin prime is an isolated prime. An isolated prime is a prime p such that $p - 2$ and $p + 2$ are not primes. Hence an isolated prime is not part of a twin prime pair.

12. For example, 23 is an isolated prime because $23 - 2$ and $23 + 2$ are not primes. The first few isolated primes are 2, 23, 37, 47, 53, 67, 79, 83, 89.

13. There are infinitely many isolated primes

Proof Consider the AP $f(n)$ defined thus: for each positive integer n,

$$f(n) = 15n + 23$$

Observe that $\gcd(15, 23) = 1$. Thus by Dirichlet's theorem, there are infinitely many primes in the AP. We seek to prove that no prime in this AP belongs to a twin prime pair. Hence we need to prove that

$$f(n) + 2 \quad \text{and} \quad f(n) - 2$$

are never primes. Now

$$\begin{aligned} f(n) + 2 &= 15n + 23 + 2 \\ &= 15n + 25 \\ &= 5(5n + 5) \end{aligned}$$

Hence, $f(n) + 25$ is composite

$$\begin{aligned} f(n) - 2 &= 15n + 23 - 2 \\ &= 15n + 21 \\ &= 3(5n + 7) \end{aligned}$$

Hence, $f(n) - 2$ is also composite. Thus no member of the AP is part of a twin prime pair. Hence each prime in the AP is an isolated prime. Since there are infinitely many primes in the AP, it follows that there are infinitely many isolated primes.

∎

14 **Prime triplets.** Consider the triple of primes 11, 13, 17. Such a list is called a prime triplet. This example is of the form $p, p+2, p+6$ where p is prime. There are also prime triplets of the form $p, p+4, p+6$. An example is $(13, 17, 19)$.

The prime triplet conjecture states that there are infinitely many prime triplets of the types $(p, p+2, p+6)$ and $(p, p+4, p+6)$.

15 Apart from $(3, 5, 7)$, we cannot have a prime triplet of the type $(p, p+2, p+4)$. The proof is given below.

16 Let p be a positive integer. If $p, p+2, p+4$ are primes then $p = 3$.

Proof Assume $p \neq 3$. Since p is prime, its only divisor greater than 1 is p. Hence $3 \nmid p$. Thus p is not of the form $3n$. Hence p is of the form $3n + 1$ or

$3n - 1$. In the first case, $p = 3i + 1$ for some integer i. Thus $p + 2 = 3i + 3$. Hence $p + 2 = 3(i + 1)$. It follows that $3 \mid p + 2$. Hence $p + 2 = 3$. Thus $p = 1$.

In the second case, $p = 3i - 1$ for some integer i. Thus $p + 4 = 3i + 3$. Hence $p + 4 = 3(i + 1)$. Thus $3 \mid p + 4$. Since $p + 4$ is prime, $p + 4 = 3$. Hence $p = -1$. Thus in each case $p = 1$ or $p = -1$. This is absurd because p is prime. Hence $p = 3$. ∎

Exercises 6.1

1 Find five ordered pairs of integers (a, b) that simultaneously satisfy the following three conditions

a $\gcd(a, b) = 1$

b $\gcd(a, b - 2) > 1$

c $\gcd(a, b + 2) > 1$

2 Use one of the pairs (a, b) to replace $(15, 23)$ in the proof that there are infinitely many isolated primes.

3 Generalize the previous exercise. Show that any ordered pair (a, b) that satisfies the three condition will form an AP $f(n) = an + b$ which contains infinitely many non-twin primes.

4 Let p be prime and $p = ab$ for some positive integers a, b. Prove that if $a > b$ then $a = p$ and $b = 1$.

5 Prove that the product of two twin primes is one less than a perfect square

6.2 Large gaps between primes

1 Twin primes have exactly one composite integer between them. The primes 7 and 11 have three consecutive integers 8, 9, 10 between them. They have a gap of $11 - 7 = 4$ between them. What is the maximum gap we can have between primes? According to the following theorem, there is no maximum gap. Thus for each positive integer n, we can find n consecutive composite integers.

2 Given any integer $n > 1$, there exist n consecutive integers, all of which are composite.

Proof We are given that $n > 1$. Consider the integers $a_1, a_2, \ldots, a_n$ defined thus

$$a_1 = 2 + (n+1)!$$
$$a_2 = 3 + (n+1)!$$
$$a_3 = 4 + (n+1)!$$
$$\vdots$$
$$a_n = (n+1) + (n+1)!$$

Now $a_1, a_2, \ldots, a_n$ are n consecutive integers. For $1 \leq r \leq n$ we have that

$$a_r = (r+1) + (n+1)!$$

We know that

$$1 \leq r \leq n$$

add 1 throughout

$$2 \leq r + 1 \leq n + 1 \quad (1)$$
$$r + 1 > 1 \quad (2)$$

since

$$(n+1)! > 0$$

add $r + 1$ to both sides

$$(r+1) + (n+1)! > r + 1 \quad (3)$$

from (2) and (3) we can write

$$1 < r + 1 < (r+1) + (n+1)! \quad (4)$$

from (1) we get

$$r + 1 \mid 1 \cdot 2 \cdots r(r+1) \cdots n(n+1)$$
$$r + 1 \mid (n+1)!$$

since $r + 1 \mid r + 1$

$$r + 1 \mid (r+1) + (n+1)! \quad (5)$$

From (4) and (5) we can conclude that $(r+1) + (n+1)!$ is composite. Thus a_r is composite for all $1 \leq r \leq n$. Hence $a_1, a_2, \ldots, a_n$ are composite. Since they are consecutive then we have n consecutive composite integers.

∎

6.3 Sequence of prime gaps

1. In this section, we define the sequence of prime gaps. We discuss the parity and size of prime gaps. Then we discuss some theorems and conjectures about the sequence of prime gaps.

2. **Definition.** For all positive integers n, let p_n be the nth prime.

3. The sequence
$$p_2 - p_1, p_3 - p_2, p_4 - p_3, \ldots$$
is the sequence of prime gaps. The first few terms of the sequence are
$$1, 2, 2, 4, 2, 4, 2.$$

4. **Parity of prime gaps.** The gaps between primes is always even except for the primes 2 and 3.

5. This is because 2 is the only even prime. All other primes are odd. Since the difference of two odd numbers is even, it follows that the gap is always even after the prime 2. We formalize and prove this statement below.

6. For all integers $n > 2$, $p_n - p_{n-1}$ is even

    Proof Let n be an integer greater than 2. Thus $n - 1 > 1$. Hence $p_{n-1} > p_1$. Thus p_{n-1} is a prime greater than 2. Thus p_{n-1} and p_n are odd primes. Since p_{n-1} and p_n are odd, their difference $p_n - p_{n-1}$ is even. ∎

7. **Size of prime gaps.** Previously, we proved a theorem about large gaps of primes. From that theorem, we know that there is no limit to the size of prime gaps. In fact, for each positive integer n, there exists an integer greater than n in the sequence. Thus the sequence has no greatest term.

8. **Theorems and conjectures about prime gaps.** The sequence of prime gaps has no obvious pattern. However there are some theorems and conjectures that describe its behaviour.

9. When three or more consecutive terms of the sequence are equal, the primes are in AP. According to Green-Tao theorem, for each positive integer n, we can find somewhere in the sequence where there is more than n consecutive terms that are equal. For example, we can find somewhere in the sequence where there are more than 100 consecutive terms that are equal.

10 The Polymath project proved that there exists an integer less than or equal to 246 that occurs infinitely many times in the sequence. Since all terms in the sequence apart from the first term are even positive integers it follows that at least one of the integers $2, 4, 6, \ldots, 246$ occurs infinitely many times in the sequence.

11 It is conjectured that every even positive integer is in the sequence of prime gaps. The twin prime conjecture holds if and only if the integer 2 occurs infinitely many times in the sequence.

Exercises 6.3

Let $1, 2, 2, \ldots$ be the sequence of prime gaps. Show that consecutive 2's only appear once in the sequence.

6.4 Change in length of gap

1 Let us study how prime gaps change in size. Sometimes the prime gap increases. For example, consider the primes p_1, p_2, p_3. We know that

$$p_2 - p_1 = 3 - 2 = 1$$
$$p_3 - p_2 = 5 - 3 = 2$$

Thus

$$p_3 - p_2 > p_2 - p_1$$

Hence the prime gap increased.

2 Sometimes the prime gap decreases. For example, consider the primes p_4, p_5, p_6. We have that

$$p_5 - p_4 = 11 - 7 = 4$$
$$p_6 - p_5 = 13 - 11 = 2$$

Thus $p_6 - p_5 < p_5 - p_4$

Chapter 6 Prime gaps

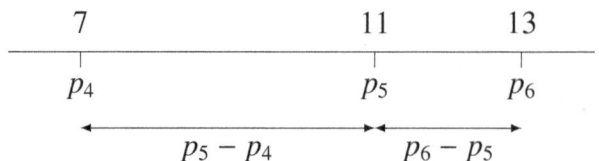

Hence the prime gap decreased.

3 Sometimes the prime gap does not change. For example, consider the primes p_2, p_3, p_4. We have that

$$p_3 - p_2 = 5 - 3 = 2$$
$$p_4 - p_3 = 7 - 5 = 2$$

Thus $p_4 - p_3 = p_3 - p_2$

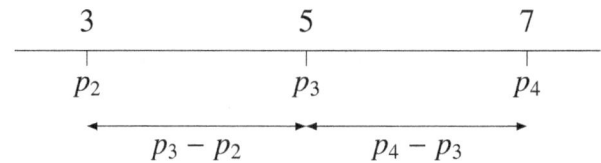

Hence the prime gap did not change.

4 It has been proven that the prime gaps are equal infinitely many times. It has also been proven that prime gaps increase infinitely many times and decrease infinitely many times.

CHAPTER 7

Distribution of primes

7.1 Introduction

1. The distribution of primes refers to how the primes are arranged among the positive integers. There are several theorems and functions that help us understand the distribution of primes. We treat some of them in this chapter. They include

   1. Bertrand's postulate
   2. prime-counting function
   3. prime number theorem

2. In this section, we show that the primes have a distribution that is not like many other sequences in number theory.

3. Let us compare these three sequences: even positive integers, squares and primes.

4. Even positive integers: $2, 4, 6, 8, 10, 12, \ldots$

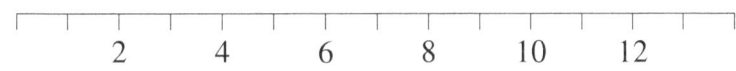

Squares: $1, 4, 9, 16, 25, 36, \ldots$

Primes: $2, 3, 5, 7, 11, 13, \ldots$

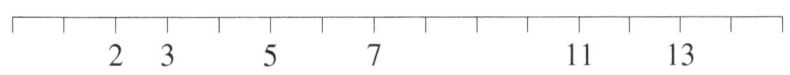

5 Observe that the even positive integers are evenly spaced. The squares have spaces that are increasing. However the primes have spaces that seem unpredictable. The spaces between the primes increases sometimes and decreases other times.

6 Let n be a positive integer. Then p_{n-1}, p_n and p_{n+1} are three consecutive primes. Thus $p_n - p_{n-1}$ is the gap between p_n and p_{n+1}.

If the gap increases we have
$$p_{n+1} - p_n > p_n - p_{n-1}$$
If the gap decreases we have
$$p_{n+1} - p_n < p_n - p_{n-1}$$

7 In the sequence of primes we can find many instances where the gap increases. On the other hand we can find many instances where the gap decreases. At other times the gap does not change. In this case the primes p_{n-1}, p_n, p_{n+1} are in AP.

7.2 Bertrand's postulate

1 Let n be an integer greater than 1. Bertrand's postulate states that there exists a prime p such that $n < p < 2n$.

2 This theorem was first proven by Chebyshev in 1852. The proof is beyond the scope of this book.

3 **Using Bertrand's postulate to prove inequalities.** We can use Bertrand's postulate to improve our previous inequalities about prime numbers. Let $p_1, p_2, \ldots$ be the sequence of prime numbers. Previously we showed that for all integers $n > 2$,
$$p_n < p_1 \cdots p_{n-1} - 1$$

4 Using Bertrand's postulate, we show that for all integers n,
$$p_n < p_1 + \cdots + p_{n-1} \quad \text{if } n > 3$$

Proof We prove the statement by induction. For the base case, let $n = 4$. We know that $p_4 = 7$ and $p_1 + p_2 + p_3 = 2 + 3 + 5 = 10$. Hence $p_4 < p_1 + p_2 + p_3$. Thus the base case is true. For the induction step, assume the statement is true for $n = k$. Thus
$$p_k < p_1 + \cdots + p_{k-1} \tag{1}$$

7.2 Bertrand's postulate

Due to Bertrand's postulate, there exists a prime p such that $p_k < p < 2p_k$. We know that p_{k+1} is the least prime greater than p_k. Since $p > p_k$, it follows that $p \geq p_{k+1}$. Hence $2p_k > p$ and $p \geq p_{k+1}$. It follows that $2p_k > p_{k+1}$. Add p_k to both sides of (1)

$$p_k + p_k < p_1 + \cdots + p_{k-1} + p_k$$
$$2p_k < p_1 + \cdots + p_k$$

Since $p_{k+1} < 2p_k$, we can write

$$p_{k+1} < p_1 + \cdots + p_k$$

Thus the statement is true at $n = k + 1$. This completes the induction step. Therefore the statement is true for all integers $n > 3$. ∎

5 For all integers n greater than 1, $p_n < 2p_{n-1}$

Proof Let n be an integer greater than 1. Thus $n - 1$ is a positive integer. Hence p_{n-1} is an integer greater than 1. By Bertrand's postulate, there exists a prime p such that $p_{n-1} < p < 2p_{n-1}$. Since $p > p_{n-1}$, it follows that $p = p_n$ or $p > p_n$. Hence $p \geq p_n$. Thus $2p_{n-1} > p$ and $p \geq p_n$. Hence $2p_{n-1} > p_n$. Thus $p_n < 2p_{n-1}$ ∎

6 For all integers n greater than 1, $p_n < 2^n$

Proof We prove this theorem by induction. For the base case, let $n = 2$. Observe that $p_2 = 3$ and $2^2 = 4$. Thus $p_2 < 2^2$. Hence the base case is established.

For the induction step, assume the statement is true at $n = k$. Thus

$$p_k < 2^k$$

We know that there exists a prime p such that $2^k < p < 2 \cdot 2^k$. Since $p > 2^k$ and $2^k > p_k$, it follows that $p > p_k$. Hence $p \geq p_{k+1}$. Thus $2 \cdot 2^k > p$ and $p \geq p_{k+1}$. Hence, $2 \cdot 2^k > p_{k+1}$. Thus $2^{k+1} > p_{k+1}$. Hence $p_{k+1} < 2^{k+1}$. Thus the statement is true at $n = k + 1$. This establishes the induction step. Therefore the statement is true for all integers n greater than 1. ∎

7 For all integers n,

$$p_n \geq p_{n-1} + 2 \quad \text{if } n > 2$$

Proof The primes $p_2, p_3, \ldots$ are all odd. Hence $p_n - p_{n-1}$ is an even integer for $n > 2$. Since $p_n > p_{n-1}$ it follows that $p_n - p_{n-1} > 0$. Hence $p_n - p_{n-1}$ is an even positive integer. The least even positive integer is 2. Thus $p_n - p_{n-1} \geq 2$. Therefore,

$$p_n \geq p_{n-1} + 2 \quad \text{if } n > 2$$

∎

8 For all integers n, if $n > 0$ then

$$p_n \geq 2n - 1$$

Proof We prove the statement by induction. For the base case, let $n = 1$ or $n = 2$. If $n = 1$ then $p_n = 2$. Observe that $2(1) - 1 = 1$. Thus $p_n \geq 2n - 1$ for $n = 1$. If $n = 2$ then $p_n = 3$. Observe that $2(3) - 1 = 5$. Thus $p_n \geq 2n - 1$ for $n = 2$. Thus we have established the base cases.

For the induction step, assume it is true for $n = k$, where $k > 1$. Thus

$$p_k \geq 2k - 1 \tag{1}$$

Since $k > 1$, $k + 1 > 2$. By a previous theorem,

$$p_{k+1} \geq p_k + 2$$
$$p_{k+1} - 2 \geq p_k \tag{2}$$

From (1) and (2), we get $p_{k+1} - 2 \geq p_k$ and $p_k \geq 2k - 1$. Hence

$$p_{k+1} - 2 \geq 2k - 1$$

Add 2 to both sides

$$p_{k+1} \geq 2k - 1 + 2$$
$$\geq 2(k+1) - 1$$

Thus the statement is true at $n = k + 1$. Therefore $p_n \geq 2n - 1$ for all positive integers n. ∎

9 **Factorial is not square.** Here we show that a factorial greater than 1 is not a square. In order to prove this we need some lemmas:

- Let p be the greatest prime less than or equal to n, where n is an integer greater than 1. Then $2p > n$.

7.2 Bertrand's postulate

- For each prime p, each integer that is strictly between p and $2p$ is relatively prime to p.

- Let n be an integer greater than 1. If $p^2 \nmid n$ for some prime factor p of n then n is not a perfect square.

10 Let p be the greatest prime less than or equal to n, where n is an integer greater than 1. Then $2p > n$.

Proof Assume $2p \leq n$. By Bertrand's postulate, there exists a prime q such that $p < q < 2p$. Hence $q < 2p$ and $2p \leq n$. Thus $q \leq n$. Hence q is a prime less than or equal to n and q is greater than p. This is absurd because p is the greatest prime less than or equal to n. Hence $2p > n$. ∎

11 For each prime p, each integer that is strictly between p and $2p$ is relatively prime to p.

Proof Let p be prime. Let n be an integer such that $p < n < 2p$.

Assume $\gcd(n, p) \neq 1$. Thus $\gcd(n, p) = p$. Hence $p \mid n$. Thus $n = p \cdot k$ for some integer k. Since $n, p > 0$, $\frac{n}{p} > 0$. Thus $k > 0$. Hence $k \geq 1$. Since $n \neq p$, it follows that $k \neq 1$. Thus $k > 1$

$$k \geq 2$$
$$pk \geq 2p$$
$$n \geq 2p$$

This is absurd because $n < 2p$. Thus $\gcd(n, p) = 1$. ∎

12 Let n be an integer greater than 1. If n is a perfect square then $p^2 \mid n$ for each prime factor p of n.

Proof Let p be a prime factor of n. Since n is a perfect square, it follows that $n = k^2$ for some nonnegative integer k. Since $n > 1$, it follows that $k^2 > 1^2$. Thus $k > 1$. Hence k has a prime factorization:

$$k = p_1 \cdots p_m$$

for some primes $p_1, \ldots, p_m$. Thus,

$$n = p_1^2 \cdots p_m^2 \qquad (1)$$

We know that

$$p \mid n$$
$$p \mid k^2$$
$$p \mid k$$
$$p \mid p_1 \cdots p_m$$

Hence p is in the list $p_1, \ldots, p_m$. From (1), we know that the square of each prime in $p_1, \ldots, p_m$ is a factor of n. Thus p^2 is a factor of n. ∎

13 Let n be an integer greater than 1. If $p^2 \nmid n$ for some prime factor p of n then n is not a perfect square.

Proof This is the contrapositive of the above theorem. ∎

14 If $n > 1$ then $n!$ is not a perfect square

Proof Let n be an integer greater than 1. Let p be the greatest prime not greater than n. Hence $n < 2p$ and $p \mid n!$. Each integer in $1, \ldots, p-1$ is relatively prime to p. Thus $1 \cdot 2 \cdots (p-1)$ is relatively prime to p. It follows that

$$\gcd(p, (p-1)!) = 1 \qquad (1)$$

From a previous lemma, each integer strictly between p and $2p$ is relatively prime to p. Since $n < 2p$, it follows that each integer in $p+1, \ldots, n$ is strictly between p and $2p$. Thus the integers $p+1, \ldots, n$ are relatively prime to p. Thus $(p+1) \cdots n$ is relatively prime to p. Therefore,

$$\gcd(p, (p+1) \cdots n) = 1 \qquad (2)$$

From (1) and (2) we get

$$\gcd(p, (p-1)!(p+1) \cdots n) = 1$$

Multiply both sides by p

$$p \cdot \gcd(p, (p-1)!(p+1) \cdots n) = p$$
$$\gcd(p^2, (p-1)!p(p+1) \cdots n) = p$$
$$\gcd(p^2, 1 \cdots (p-1)p(p+1) \cdots n) = p$$
$$\gcd(p^2, n!) = p$$
$$\gcd(p^2, n!) \neq p^2$$

Thus

$$p^2 \nmid n!$$

Thus there exists a prime factor p of $n!$ such that $p^2 \nmid n!$. It follows that $n!$ is not a perfect square. ∎

Exercises 7.2

1 If $n \geq 5$ then $p_n > 2n - 1$.

(Hint: Use induction.)

2 For each integer n greater than 2, there exists a prime p such that $n < p < n!$

(Hint: Let p be a prime factor of $n! - 1$.)

3 Let a, b be integers greater than 1. If $a \leq \frac{b}{2}$ then there exists a prime p such that $a < p < b$.

(Hint: Use Bertrand's postulate.)

4 Let n be an integer greater than 1. There exists a prime number p such that
$$p \leq n < 2p$$

(Hint: Let p be the greatest prime not greater than n.)

5 If p_n denotes the nth prime number then
$$p_1 p_2 \cdots p_n + 1$$
is not a perfect square

(Hint: Show that $p_1 p_2 \cdots p_n + 1$ is not of the form $4k + 1$.)

7.3 Prime-counting function

1 In this section, we define the prime-counting function and prove some of its properties.

2 **Definition.** For all positive real numbers x, we define $\pi(x)$ as the number of prime numbers not greater than x.

3 The primes not greater than 10 are 2, 3, 5, 7. Thus $\pi(10) = 4$. However, $\pi(2) = 1$ since 2 is the only prime not greater than 2.

As another example, $\pi(0.5)$ is zero because there is no prime that is not greater than 0.5. Similarly, $\pi(1) = 0$.

4 **Properties of the prime-counting function.** Here we prove the following properties of the prime-counting function.

- Let x, y be real numbers greater than or equal to 2. If $x \leq y$ then $\pi(x) \leq \pi(y)$.
- For all real numbers x, $\pi(x) = \pi(\lfloor x \rfloor)$
- For all positive integers n,
$$\pi(p_n) = n.$$
- For all integers n, if n is prime then $p_{\pi(n)} = n$
- For all real numbers x, if $x \geq 2$ then $p_{\pi(x)} \leq x$
- For all real numbers x, if $x \geq 2$ then $\pi(x) < \pi(2x)$.
- For all positive integers n,
$$\pi(2n - 1) \leq n \leq \pi(2^n)$$

5 Let x, y be real numbers greater than or equal to 2. If $x \leq y$ then $\pi(x) \leq \pi(y)$.

Proof Let p be a prime that is less than or equal to x. Hence $p \leq x$ and $x \leq y$. Thus $p \leq y$. Hence each prime that is less than or equal to x is also less than or equal to y. It follows that $\pi(x) \leq \pi(y)$. ∎

6 For all real numbers x,
$$\pi(x) = \pi(\lfloor x \rfloor)$$

Proof Let p be a prime such that $p \leq x$. Since p is an integer, it follows that $p \leq \lfloor x \rfloor$. Thus all primes less than or equal to x are also less than or equal to $\lfloor x \rfloor$. Hence
$$\pi(x) \leq \pi(\lfloor x \rfloor) \tag{1}$$
Since $\lfloor x \rfloor \leq x$, it follows that
$$\pi(\lfloor x \rfloor) \leq \pi(x) \tag{2}$$
From (1) and (2), we conclude that
$$\pi(x) = \pi(\lfloor x \rfloor)$$
∎

7 For all positive integers n,
$$\pi(p_n) = n.$$

7.3 Prime-counting function

Proof The first n primes are $p_1, \ldots, p_n$. Thus there are exactly n primes not greater than p_n. Hence

$$\pi(p_n) = n$$

∎

8 For all integers n, if n is prime then $p_{\pi(n)} = n$

Proof Since n is prime, it follows that $n = p_k$ for some positive integer k. From a previous theorem,

$$\pi(p_k) = k$$

Thus,

$$\pi(n) = k$$
$$p_{\pi(n)} = p_k$$
$$= n$$

∎

9 For all real numbers x, if $x \geq 2$ then $p_{\pi(x)} \leq x$

Proof Let x be a real number such that $x \geq 2$. Let k be the number of primes not greater than x. Thus $k = \pi(x)$. Hence p_k is among the primes not greater than x. Hence $p_k \leq x$. Therefore $p_{\pi(x)} \leq x$. ∎

10 For all real numbers x, if $x \geq 2$ then $\pi(x) < \pi(2x)$.

Proof Let x be a real number such that $x \geq 2$. Hence, $\lfloor x \rfloor \geq 2$. Let $n = \lfloor x \rfloor$. By Bertrand's postulate, there exists a prime p such that $n < p < 2n$. Thus p is a prime less than $2n$ but greater than n. Hence $2n$ has more primes less than or equal to it compared to n. Thus

$$\pi(2n) > \pi(n)$$
$$\pi(2\lfloor x \rfloor) > \pi(\lfloor x \rfloor)$$

By a previous theorem,

$$\pi(\lfloor x \rfloor) = \pi(x)$$

Thus
$$\pi(2\lfloor x \rfloor) > \pi(x)$$

Since
$$\lfloor x \rfloor \leq x$$
$$2\lfloor x \rfloor \leq 2x$$

Thus
$$\pi(2\lfloor x \rfloor) \leq \pi(2x)$$

Hence $\pi(2x) \geq \pi(2\lfloor x \rfloor)$ and $\pi(2\lfloor x \rfloor) > \pi(x)$. Thus $\pi(2x) > \pi(x)$. ∎

11 Let n be a positive integer. Then
$$\pi(2^n) \geq n$$

Proof From a previous theorem, $p_n \leq 2^n$. Thus $\pi(p_n) \leq \pi(2^n)$. Hence $n \leq \pi(2^n)$. Therefore $\pi(2^n) \geq n$. ∎

12 For all positive integers n,
$$\pi(2n - 1) \leq n$$

Proof By a previous theorem,
$$p_n \geq 2n - 1$$

Thus
$$\pi(p_n) \geq \pi(2n - 1)$$
$$n \geq \pi(2n - 1)$$
$$\pi(2n - 1) \leq n$$

∎

13 For all positive integers n,
$$\pi(2n - 1) \leq n \leq \pi(2^n)$$

Proof By previous theorems,
$$\pi(2n - 1) \leq n \quad \text{and} \quad n \leq \pi(2^n)$$

Thus
$$\pi(2n - 1) \leq n \leq \pi(2^n)$$

∎

7.5 Prime number theorem

1. In this section, we discuss the prime number theorem. Before this, we define the asymptotic relation. This is an important relation used to state the prime number theorem.

2. **The asymptotic relation ~.** Let f and g be functions. Then we write $f \sim g$ if and only if
$$\lim_{x \to \infty} \frac{f(x)}{g(x)} = 1.$$

3. We say f and g are asymptotically equivalent.

4. The relation $\sim$ is a binary relation. It has properties that make it an equivalence relation: Let f, g, h be functions. Then

    1. Reflexive. $f \sim f$
    2. Symmetric. If $f \sim g$ then $g \sim f$
    3. Transitive. If $f \sim g$ and $g \sim h$ then $f \sim h$

5. Let us consider an example of a pair of functions that are asymptotically equivalent. Consider the functions x and $x + 1$. We have that
$$x + 1 \sim x.$$

6. **Prime number theorem.** The prime number theorem is an important theorem in number theory. It states that for all real numbers x, if $x > 2$ then
$$\pi(x) \sim \frac{x}{\ln x}.$$

7. This theorem is useful because it means we can approximate the behaviour of a complicated function like $\pi(x)$ using a simpler function: $x/\ln(x)$.

8. From the prime number theorem, we can predict that the primes become less common as they become larger.

9. Exactly 40% of the positive integers not greater than 10 are prime. Exactly 25% of the positive integers not greater than 100 are prime. Exactly 16.8% of the positive integers not greater than 1000 are prime.

10. As the number gets larger, the percentage of primes not greater than it gets smaller

Exercises 7.3

Prove the following.

1. Let n be an integer greater than 1.

   a. $\pi(n) \leq \pi(n+1)$

   b. $\pi(n) \leq \pi(n-1) + 1$

   c. $\pi(n) > \pi(n-1)$ if and only if n is a prime number.

   d. $\pi(n) = \pi(n-1) + 1$ if and only if n is a prime number.

   e. n is composite if and only if $\pi(n) = \pi(n-1)$.

2. Let m, n be integers greater than 1.

   a. If $m < n$ then $\pi(n) - \pi(m)$ is the number of primes p such that $m < p \leq n$.

   b. If $m \leq n$ then $\pi(n) - \pi(m-1)$ is the number of primes p such that $m \leq p \leq n$.

3. Let p_n be the nth prime number for each positive integer n. Then $\pi(p_n + 1) = n$ if $n > 1$.

4. Let n, k be positive integers. If $k+1, k+2, \ldots, k+n$ are composite integers then
$$\pi(k+1) = \pi(k+2) = \cdots = \pi(k+n)$$

7.4 Scarcity of primes

1. To measure the scarcity of primes we count how many primes are less than or equal to a given real number.

2. The prime-counting function π is used for this purpose.

3. Consider the 100 consecutive integers $1, \ldots, 100$. We know that $\pi(100) = 25$. Thus the probability of an integer in the list being prime is $\frac{25}{100} = 25\%$.

4. Consider the 100 consecutive integers $101, 102, \ldots, 200$. Since $\pi(200) - \pi(100) = 21$, it follows that there are 21 primes in the list $101, \ldots, 200$. Hence the probability of an integer in the list being prime is 21%.

5. Thus the probability decreased from 25% to 21%. This trend continues as the numbers get larger. Hence the primes become scarcer as we move along the number line.

www.ingramcontent.com/pod-product-compliance
Lightning Source LLC
Chambersburg PA
CBHW062115220526
45471CB00010B/3742